Entomology
Principles and Practices

Entomology
Principles and Practices

H. H. Robinson

AGROBIOS (INDIA)

Published By:
AGROBIOS (INDIA)
Agro House, Dehind Nasrani Cinema
Chopsani Road, Jodhpur - 342 003
Phone: +91291-2643993, 2643994, Fax: 2642319
E. Mail: agrobiosindia@gmai;.com
Website: agrobiosindia.com

AGROBIOS (INDIA)

2012

9 788177 541090
Rs. 1995.00 / US$ .00
Published by:
AGROBIOS (INDIA)

Published by: Dr. Updesh Purohit for Agrobios (India), Jodhpur
Laser Typesetting: Yashee Computers, Jodhpur
Cover Design by: Reena
Printed at: Bharat Printers (Press), Jodhpur

PREFACE

THROUGHOUT the past century an increasing amount of scientific research work has been carried out in connection with the various branches of Agriculture and Horticulture. Not only have problems such as soil fertility and plant-breeding engaged the attention of many workers, but it has also become apparent that crops must be protected in some way against fungus diseases and injurious insects.

Little interest appears to have been taken in injurious farm insects until Curtis, in 1841, published his first contribution on the subject in the *Journal of the Royal Agricultural Society of England*. He subsequently published further papers, and in 1881 Miss Ormerod published a Manual of Injurious Insects. From that time onwards the subject made great strides, until nowadays a working knowledge of the subject of Entomology is regarded as part of a farmer's equipment.

It is impossible to devise, or even appreciate, measures which can be taken to prevent or counteract insect attacks, unless the essentials of the life history and habits of the insect concerned are understood. Part I. of this book deals, therefore, with the elements of Entomology in as concise a form as possible, in order to give the student a grasp of this part of the subject. Part II. deals, under their respective orders, with those individual insects which are important pests, and where methods of control are already known, these are given. A separate chapter deals with the general principles underlying the control of insect damage. On a small scale, or with fruit pests, it is often worth while to take direct measures in the form of applications of insecticidal substances. Usually, however, this is not possible on the farm, and therefore every effort has been made to show how various other operations, *e.g.*, a variation in rotation, and cultural methods, can achieve the desired end.

A number of animals, some of them closely allied to insects, are responsible for damage to crops and stock, the symptoms of their attack being very similar to those occasioned by true insects. These animals are dealt with in the Appendices to this book. The classification followed in Appendix I. is that set out by the Natural History Department of the British Museum, in the "Guides."

Professor F. J. Cole, F.R.S., and Dr N. B. Eales, of the University of Reading, very kindly read through and criticised the MS. in its early stages. We are indebted to them for several valuable suggestions. Our colleague Mr E. L. Crossley kindly undertook the task of reading the proofs.

Of the illustrations, Figs. 1, 2, 4, 5, 6, 8, 9, 10, 13, 14, 15, 18, 19, 23, 24, 25, 32, 34, 51, 101, 102, 104, are original drawings by the authors from fresh specimens.

The Trustees of the British Museum (Natural History) allowed us to reproduce, from their Guides, Figs. 12, 26, 29, 31, 108, 132: they also gave us permission to copy several illustrations, namely, Figs. 139, 140, 141.

The Imperial Bureau of Entomology provided the block for Fig. 28 from *The Bulletin of Entomological Research.*

The Ministry of Agriculture and Fisheries and H.M. Stationery Office provided the blocks for Figs. 27, 73, 135. These had previously appeared in *The Journal of the Ministry of Agriculture.*

The Editor of the *Annals of Applied Biology* and the Cambridge University Press supplied blocks for Figs. 20, 33, 45, 79, 82, 97, 98, 100.

The Cambridge University Press also supplied electros for the following illustrations. From *Typical Flies*, by E. K. Pearce, Figs. 106, 109, 111, 112, 114, 115, 117, 120, 122, 123, 125, 126, 127. From *Insect Pests and Fungus Diseases*, by P. J. Fryer, Figs. 21, 55, 58, 60, 62, 67, 80.

Messrs Edward Arnold & Co. supplied blocks from *British Hymenoptera*, by Buckhurst, Staniland & Watson, for Figs. 64, 87, 89, 90, 91, 92, 93, 95, 96. Fig. 88 was copied from this work.

Messrs Methuen & Co. supplied us with electros of Figs. 16, 36, 133, from *A General Text Book of Entomology*, by A. D.

Imms. Figs. 3, 17, 37, 47, 48, 74, have been re-drawn from figures in this book. Messrs Methuen also allowed us to reproduce Figs. 42, 99, 103, 107, from *British Insects*, by Harold Bastin.

Messrs A. & C. Black supplied the electro for Fig. 68, and the Oxford Medical Publications that for Fig. 144, *ex* Thomson's *Outlines of Zoology*.

Messrs W. J. Craven & Co., Ltd., supplied a photograph for Fig. 56.

Professor F. V. Theobald, of the South Eastern Agricultural College, allowed us to use Figs. 38, 39, 63, 74A, 94, 129, from his publications, the photographs being by F. Edenden. Fig. 137 is a copy of one of Theobald's figures.

Professor R. S. MacDougall, of the University of Edinburgh, and Messrs W. Blackwood & Sons, Ltd., allowed us the use of Figs. 40, 41, 57, 59, from the *Journal of the Highland and Agricultural Society*.

Mr H. W. Miles supplied us with fresh prints of Figs. 44 and 61, which originally appeared in the *Journal of the Bath and West Society*. He also took photographs of material in our possession for Figs. 46, 116, 118, 121, and gave us the print for Fig. 148. Fig. 131 is from a photograph by Prof. W. T. Wilson. Figs. 43 and 105 have been re-drawn from several sources.

Figs. 22, 30, 49, 50, 52, 53, 54, 65, 66, 69, 70, 71, 72, 75, 76, 77, 78, 81, 83, 84, 85, 86, 110, 142, 143, are copyright photos by Mr A. Harold Bastin, some of which have already appeared in the *Journal of the Bath and West Society* and elsewhere, whilst many were specially arranged and taken for this book.

Figs. 145, 146, 147 are copied from Marcinowski, Arb. K., biol. Anst. Land-, Forstw., VII. Bd., 1909.

To all those who have assisted us in any way by criticisms, suggestions, or permission to make use of illustrations, we offer our sincere thanks.

D. H. R.
S. G. J.

HARPER ADAMS AGRICULTURAL COLLEGE
NEWPORT, SHROPSHIRE
November 1928

CONTENTS

PART I

GENERAL MORPHOLOGY, PHYSIOLOGY AND CLASSIFICATION

PART II

ECONOMIC SECTION

APPENDIX I

OTHER ARTHROPODA OF ECONOMIC IMPORTANCE

APPENDIX II

APPENDIX III

Part I

GENERAL MORPHOLOGY, PHYSIOLOGY AND CLASSIFICATION

CHAPTER I

GENERAL EXTERNAL CHARACTERS

THE study of Entomology is best approached through the examination of the structure and life histories of typical insects. The common Wasp, or better still, a Queen Wasp, serves as a useful type upon which to commence observations on the external structure. The Cockroach, owing to the ease with which it can be dissected, is a more suitable insect for the study of the internal structure, whilst the Capsid Bug and the Cabbage White Butterfly admirably demonstrate different kinds of life history. The first part of this book is devoted to a description of these types, together with such references to other insects as may be necessary to elucidate details of structure and life history not demonstrated by the four insects mentioned.

The Wasp.—When examining a wasp it is convenient to have the specimen mounted upon a fine pin thrust through the body between the wings. Then, with the creature in the position assumed during flight or whilst resting, the upper side can be referred to as the *dorsal* surface, and the under side as the *ventral* surface.

A glance at the dorsal surface of a wasp (Fig. 1) is sufficient to demonstrate that its body is divided into three quite distinct regions, the *head* (*H.*), *thorax* (*Th.*) and *abdomen* (*Abd.*). Each region is joined to its neighbour by a narrow isthmus of tissue. Thus the head is connected with the thorax[1] by the 'neck,' and the thorax with the abdomen by the 'waist.'

This threefold division of the trunk is characteristic, the term 'insect' having been derived from the Latin word *Insecare=to cut into,* i.e. notched.

Further examination shows that the wasp possesses three pairs of jointed legs (L_1, L_2, L_3), all of which are attached to the ventral surface of the thorax. This is another

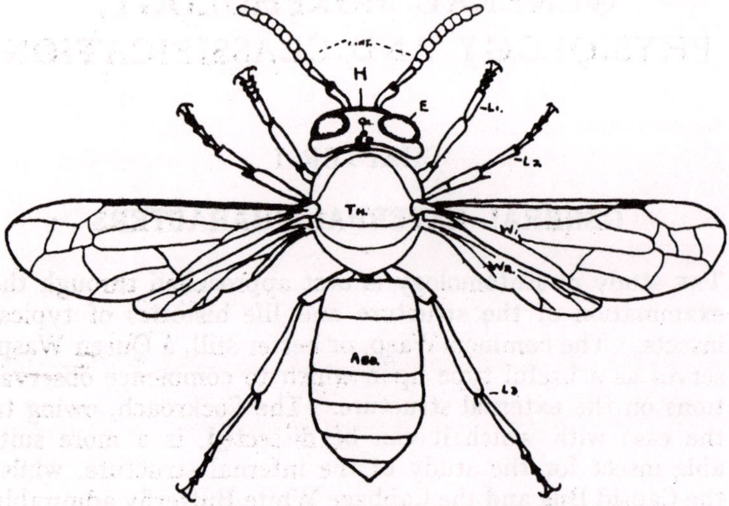

FIG. 1.—Dorsal surface of Queen Wasp.
H. Head : *Th.* Thorax : *Abd.* Abdomen : *Ant.* Antennæ : *E.* Compound Eye : *Oc.* Ocelli : L_1, L_2, L_3 Legs : W_1, W_2 Wings.

characteristic of adult insects, and one which serves to distinguish them from all other groups of animals.

This feature has led to the classification of insects as *Hexapoda* (Gr. *Hex=six : pous=a foot*), or animals with six legs.

The thorax of the wasp also bears two pairs of wings (W_1, W_2), which are attached to the dorsal surface. Most insects have wings of some sort, though they are often very different in shape, size and construction from those

[1] In the wasp, as in all Hymenopterous insects, the central region of the body consists of the thorax and the first abdominal segment fused together. In the majority of insects the thorax is separated by the 'waist' from the abdomen (see Fig. 8).

of the wasp. Some insects, such as the true Flies, have only one pair of wings: a number of degenerate insects, such as the various forms of Lice, possess no wings at all.

The posterior end of the abdomen is furnished with an apparatus for egg-laying called an *ovipositor*, which will be described later.

In the Worker Bee the ovipositor is modified into a stinging organ. In a number of insects there are *caudal* or tail appendages called *cerci*: in the Bristletails (Fig. 7) and Mayflies they are long and narrow, whilst in the Earwigs, they take the form of nippers or pincers.

Structure of the Head.—If the head of the wasp be examined from the front, certain structures become apparent (Fig. 2). The head itself is heart-shaped and flattened, with its apex towards the ventral surface. The right and left sides of this heart-shaped mass are almost entirely occupied by two large oval structures, the *compound eyes* (*E.*). On the very top of the head are three small spherical bodies arranged in a triangle. These are the simple eyes, or *ocelli* (*Oc.*). Springing from the front of the head, between the compound eyes, arise a single pair of long-jointed 'feelers' or *antennæ* (Fig. 1, *Ant.*). Towards the under side of the head is the *clypeus* (*Cl.*), which partly overlies the two large jaws or *mandibles* (*Man.*).

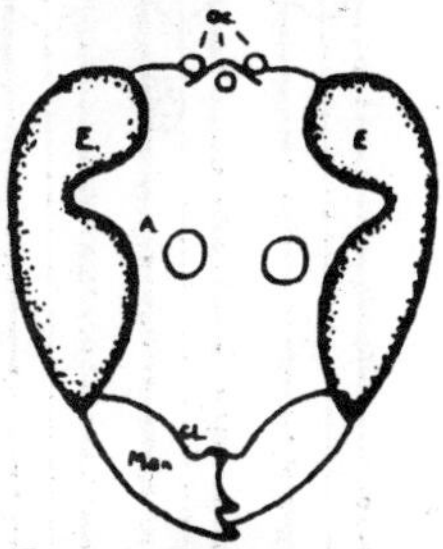

FIG. 2.—Frontal view of Head of Queen Wasp. *E.* Compound eyes: *Oc.* Ocelli: *A.* Point of origin of antennæ: *Cl.* Clypeus: *Man.* Mandible.

The Eye.—The compound eyes are often the most conspicuous part of the head. Examination with a lens or low-power microscope reveals the fact that the eye is composed of large numbers of small, polygonal lenses. Each lens is called a *facet*, and the compound eye is consequently a faceted eye. The number of facets varies with the type of the insect. In some cases there may be only from 6 to 9 facets, whilst in the eye of the Dragon-fly there may be over 28,000.

The eye is a very complicated structure, and is connected by nerves to the 'brain.' Each element or *ommatidium* consists of a minute tube with a lens at the top. The outside of each tube is coated with a dark pigment which isolates it from its neighbours. The way in which these tubes are arranged is represented diagrammatically in Fig. 3.

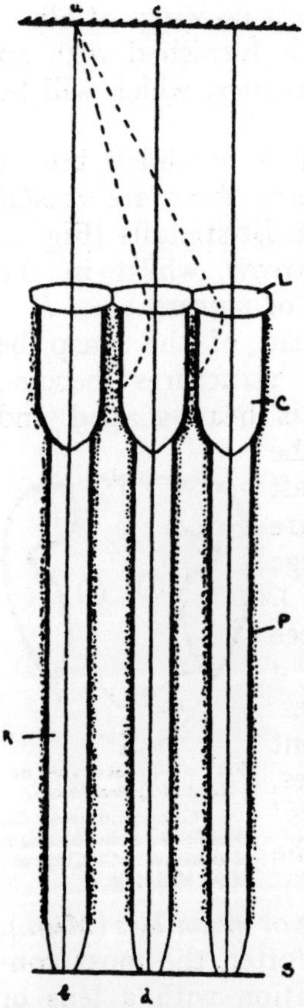

FIG. 3.—Diagrammatic longitudinal section through three elements of a compound eye.

L. Corneal lens: *C.* Crystalline cone: *P.* Pigment covering cones and retinulæ: *R.* Retinula: *S.* Sensitive layer.

Modified from Imms, after Exner.

Owing to the construction of the lens (*L*) and the presence of the pigment (*P*) on the walls of the tube, the only ray of light which can pass through a tube to form an image upon the sensitive layer (*S*) is the one parallel to the axis of the tube. Thus an image of the object *a* will be cast on the sensitive layer at *b*, because the ray of light *ab* is parallel to the axis of the left-hand tube. But no image of *a* will be cast on the sensitive layer at the base of the centre and right-hand tubes. Rays of light from *a* which strikes the lenses of these tubes will be deflected by the lenses, and will be absorbed by the pigment.

Consequently the image cast by a faceted eye is a mosaic made up of numerous points of light, each point corresponding to some part of the object seen; these points, in the aggregate, form a single coherent picture. The clearness with which an object is perceived by a compound eye depends upon the number and size of the individual facets. The smaller the facets and the greater their number, the more distinct the vision. Compound eyes are especially well adapted for perceiving movement.

The simple eye, or *ocellus*, consists of a single lens only. It gives a very indistinct image, and its chief function is distinguishing between light and darkness.

The Antennæ.—The antennæ (Fig. 1, *Ant.*) of the wasp

are slightly longer than the thorax. Since both the antennæ and the head upon which they are situated are capable of movement, they have a wide range. The antennæ in the wasp are club-shaped, consisting of numerous rounded joints at the apex and rather longer joints at the point of attachment to the head.

FIG. 4.—Common Types of Antennæ.

1. Clavate. 2. Pectinate. 3. Lamellate. 4. Geniculate. 5. Plumose.

Authorities are agreed that the antennæ are highly-developed sense organs. Both the olfactory sense, or sense of smell, and the sense of touch reside in the tip of the antennæ.

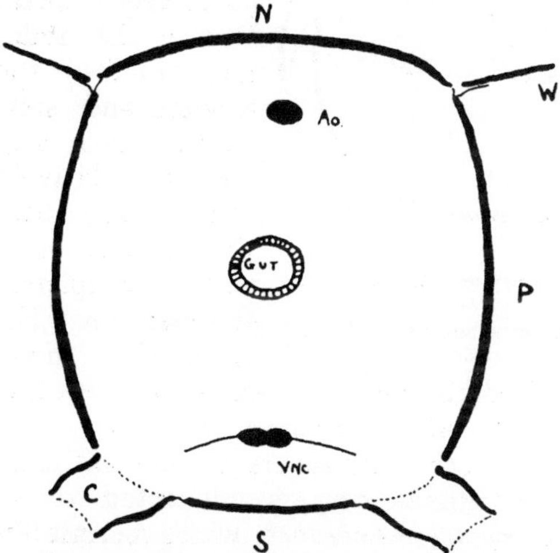

FIG. 5.—Diagrammatic transverse section through thorax of wasp.

N. Tergum : *S.* Sternum : *P.* Pleuron : *C.* Coxa : *Ao.* Aorta : *W.* Wing : *VNC.* Ventral nerve cord.

Antennæ differ widely in the various groups of insects (Figs. 4 and 68), and in some cases prove useful means of distinguishing the sexes.

Segments of the Body.—The thorax and the abdomen are divided up transversely into a number of rings; each ring is called a *segment*. Every insect shows a certain amount of segmentation, though the dividing lines between the various segments are often difficult to perceive, owing to fusion of adjacent segments.

Each body segment consists of thin plates of a tough,

hard material called *chitin*, which enclose the internal organs of the segment. There are four regions, each composed of a varying number of these plates or *sclerites* (see Fig. 5). The dorsal region (*N*) is called the *tergum*, the ventral region (*S*) the *sternum*, whilst each of the two laterals (*P*) is known as a *pleuron*.

The relative sizes of these plates vary very much in different insects, and also in the various parts of the body of any individual insect. For example, examination of the thorax of the wasp (Fig. 5) reveals the existence of well-defined dorsal and lateral plates (*tergites* and *pleurites*), whilst the ventral plates (*sternites*) are comparatively small. But in the abdomen (Fig. 6) the tergum and sternum of each segment are relatively enormous, and they overlap so as to hide completely the pleurites. The pleurites, moreover, are not chitinized to the same degree as the other sclerites, but remain somewhat membranous.

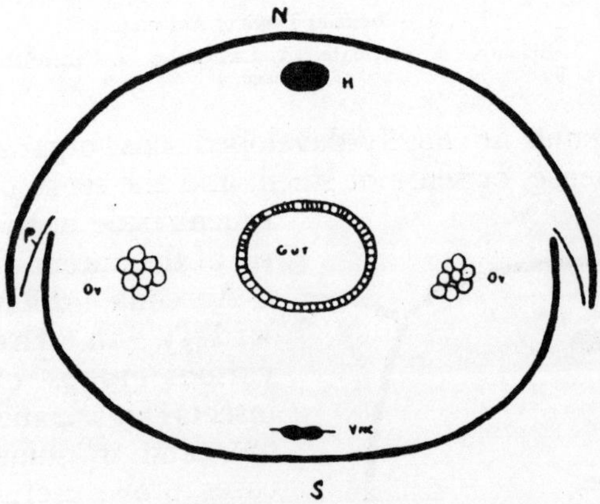

FIG. 6.—Diagrammatic transverse section through abdomen of a wasp.
Ov. Position of ovaries. Other lettering as for Fig. 5.

The sclerites are connected by layers of non-chitinized membrane. Each abdominal ring is connected to its neighbour by *intersegmental membranes*, which permit the segments to move independently. The external covering of an insect may, in fact, be compared to the steel armour worn in mediæval times. The chitinous areas correspond to the plates of armour, whilst the intersegmental membranes form the equivalent of the leather hinges.

The hard outer covering, taken as a whole, is called the *cuticle*. It serves as a point of attachment for the muscles and internal organs of the body. Insects have no internal, bony skeleton such as the higher animals possess. The chitinous cuticle performs the duties of a skeleton, and insects are said, therefore, to have an *exoskeleton* in contrast to the *endoskeleton* of vertebrate animals.

Chitin is a substance of high molecular weight and is composed of varying proportions of carbon, hydrogen, oxygen and nitrogen. It is very resistant to the action of acids and alkalies.

The bodies of insects are frequently covered with hairs of various kinds. Some of these arise directly from the cuticle, of which they are merely outgrowths; others originate in the layers under the cuticle, and pass through it to the exterior (*setæ*).

CHAPTER II

THE THORAX

THE thorax is a very complicated structure, and is a very important one, since it bears the organs of locomotion or movement. It consists of three segments, and although in theory these segments are distinct, yet in the majority of insects it is often extremely difficult to distinguish them.

In primitive insects like the bristletails the three divisions of the thorax can be readily observed (Fig. 7). The segment next the head is the *prothorax*, the middle one the *mesothorax*, and the one adjacent to the abdomen is the *metathorax*. In many insects the mesothorax extends forward over the prothorax and posteriorly over the metathorax, partially or even completely hiding these segments.

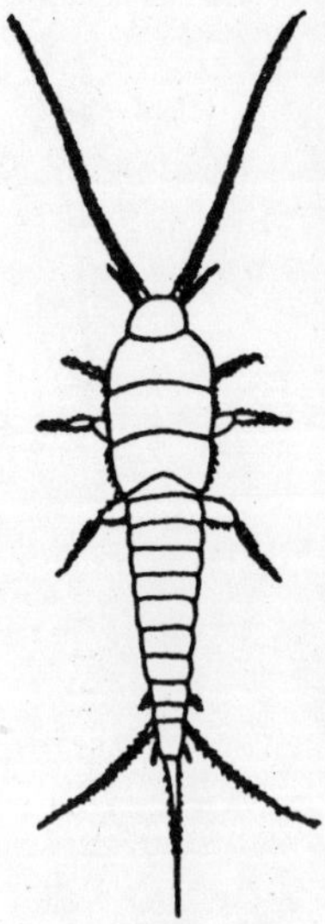

FIG. 7.—Fish Insect (magnified) showing distinct thoracic segments.

Reduced from a drawing by Lubbock.

In the wasp, almost the whole of the dorsal surface is occupied by the chitinous plates belonging to the mesothorax (Fig. 8, *B*). The prothorax forms a ring of chitin in advance of the mesothorax distinguished by its yellow margin: the metathorax is only just visible. In the wasp the problem of identification is made still more difficult because the first abdominal segment is fused with the metathorax. The 'waist' of the wasp, therefore, is not really a division between thorax and the abdomen, but divides the first abdominal segment from the posterior ones. From the elementary point of view, however, this distinction is of no importance: in the majority of insects the 'waist' does separate thorax and abdomen.

Fig. 8 *A* represents a somewhat diagrammatic view of the thorax of a wasp seen from the side. The ***mesoscutum*** (*Meso. Sc.*) or anterior dorsal plate of the mesothorax overlies the prothorax (*Pro.*), whilst the ***scutellum*** (*Scut.*) almost covers the metathorax (*Met.*). The mesothorax is constricted in the centre, but expands into the ***meso-pleuron*** (*Meso. Pl.*), giving the segment as a whole a saddle-shaped appearance.

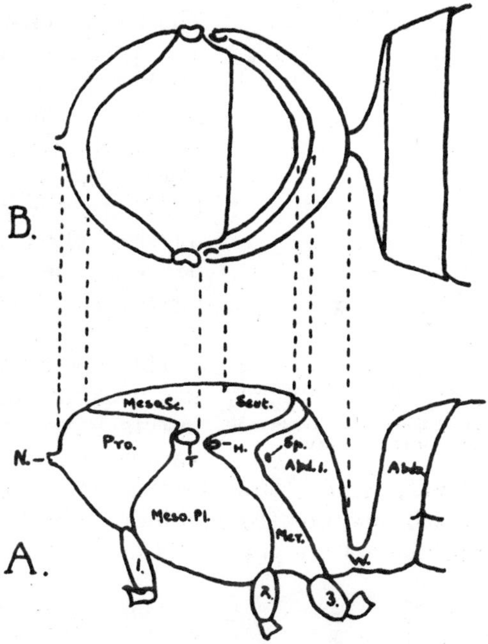

FIG. 8.—Thorax of Wasp.

A. Lateral view. *B.* Dorsal view.
N. Neck: *W.* Waist: *Pro.* Prothorax: *Meso. Sc.* Mesoscutum: *Scut.* Mesoscutellum: *Meso. Pl.* Mesopleuron: *Met.* Metathorax: *T.* Tegula: *H.* Point of attachment of hind wing: *Abd.* 1, *Abd.* 2, first and second abdominal segments: *Sp.* Spiracle: 1, 2, 3, Legs.

The pleurites of the wing-bearing segments are divided into two regions by a suture which extends from wing base to *coxal cavity*. The anterior plate is called the ***episternum*** and the posterior plate the ***epimeron***. In some insects the prothoracic pleurites are similarly divided. The dorsal plate of the prothorax is known as the ***pronotum***.

The Legs (1, 2, 3) arise on the ventral surface, one pair to each segment, each leg arising at the posterior margin of the segment. Owing to the width of the mesothorax there is consequently a considerable gap between the first and second pairs; the second and third pairs arise close together.

The legs of the wasp are all of the same general construction, though they may differ in detail (Fig. 9). That part of the leg attached to the thorax is called the *coxa*, and the other parts in their respective order are *trochanter*, *femur*, *tibia* and *tarsus*. The tarsus consists of several joints. The proximal one, or one nearest the body, is comparatively long and is termed *metatarsus*. At the distal end of the tibia are two long *tibial spurs*: the last joint of the tarsus terminates in two sharp claws, or *ungues*. Often there are small pads between the claws which enable insects to cling to smooth walls, etc.

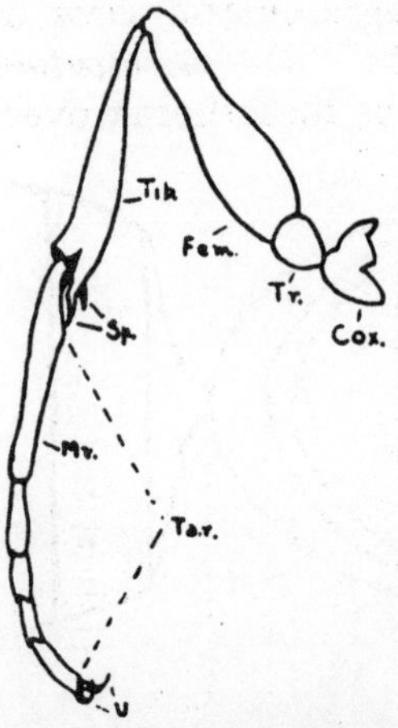

FIG. 9.—Hind Leg of Wasp.

Cox. Coxa: *Tr.* Trochanter: *Fem.* Femur: *Tib.* Tibia: *Sp.* Tibial spurs: *Tar.* Tarsus: *Mt.* Meta-tarsus: *U.* Ungues.

The legs of the wasp are designed for walking and running. Other insects have their legs modified in various ways to serve the purpose of burrowing (Scarab beetles), swimming (aquatic insects), or seizing prey (Dragon-flies), whilst in some butterflies the forelegs are practically non-existent.

CHAPTER III

THE WINGS

THE wings are found only on the two posterior thoracic segments. Wings are never carried on the prothorax. Owing to the small size of the metathorax, the two pairs of wings arise very close together. At the base of each fore wing is a prominent raised area, the *tegula* (Fig. 8, *T.*). The point of attachment of the hind wing (*H.*) is much less conspicuous.

The wings are connected with the thorax by the tegulæ which are conspicuous in the family to which the wasp belongs, but inconspicuous in other orders. Each tegula is a complicated joint designed to allow the wing to vibrate at a rapid rate.

The wing itself consists of upper and lower layers of fine membrane which enclose numerous hollow chitinous tubes called *nervures* or *veins*. In systematic work the study of the *neuration* or *venation* of wings is of considerable importance.

A typical wing has three *margins*, as shown in Fig. 10, which represents the fore and hind wings of a wasp. The anterior margin *AB* is called the *costa*. *BC* is the *apical* or *outer* margin, and *CD* forms the *inner* or *anal* margin. At *A* is the *humeral* angle, at *B* the *apex*, and at *C* the *anal* angle.

According to Comstock the wing has eight principal veins, shown in Fig. 11, which demonstrates the theoretical neuration. These are *Costa* (*C.*), *Sub-Costa* (*Sc.*), *Radius* (*R.*), *Media* (*M.*), *Cubitus* (*Cu.*), *First Anal* (*1st A.*), *Second Anal* (*2d A.*) and *Third Anal* (*3d A.*). The second, third, fourth and fifth veins are branched, and the branches are indicated by placing a number after the abbreviation for the name of the wing. Thus *R1* denotes the first branch of the Radius, and so on.

In many insects, such as the Hymenoptera, the number of veins in the wing is less than it is in the theoretical wing. In others, such as Neuroptera, the number is greater. Some of the main veins coalesce in other types and it becomes difficult to decide what any particular vein may be.

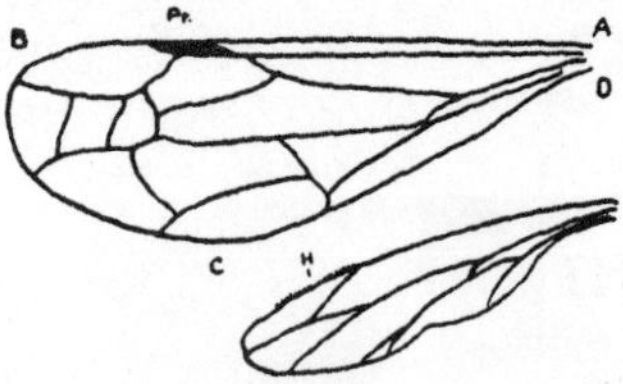

FIG. 10.—Fore and Hind Wings of Wasp.
For explanation see text.

Other authorities state that the hypothetical wing contains nine principal veins. In fact there is so much confusion of terms used in describing venation that the student is compelled to study each order of insects separately; the names given to the veins of any particular order often do not apply to the veins of any other order.

In hymenopterous insects, such as wasps, bees, etc., there is a dark area or spot underneath the costa. This is known as the *stigma* or *pterostigma* (Fig. 10, *Pt.*).

The spaces enclosed by the nervures of the wing are called *cells*. A *closed* cell is completely surrounded by veins; an *open* cell extends to the margin. Usually the nervures enclose a *trachea* (see p. 24), and sometimes a fine nerve fibre.

A number of insects are able to couple together the fore and hind wings of the right and left sides respectively during flight, thus ensuring unison of wing beats. In the Wasp and many other Hymenoptera coupling is effected by *hamuli*, or small chitinized hooks found on the anterior margin of the fore wings (Fig. 10, *H.*). Some moths, such as the Death's Head Moth, have a projecting bristle or *frenulum* on the costa of the hind wing, which fits under a strap-like structure called the *retinaculum*, carried on the

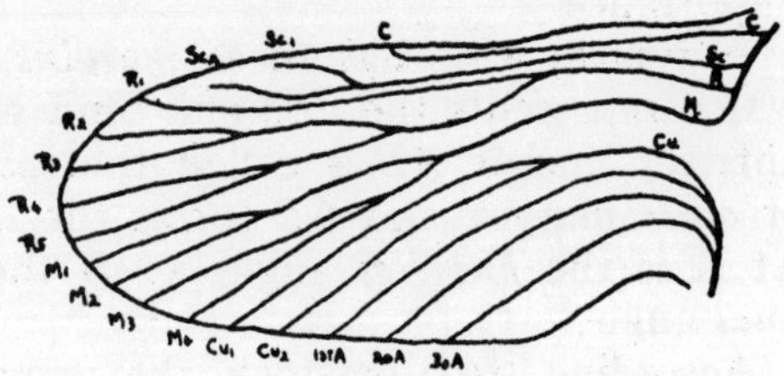

FIG. 11.—Theoretical Venation of Wing. After Comstock.
For explanation see text.

under side of the fore wing (Fig. 12). In the Noctuids both frenulum and retinaculum are composed of bristles.

Wing modifications are common. The fore wings of beetles are strongly chitinized, forming wing covers or *elytra*. They take no active part in flying, but are held rigid, away from the body and out of the way of the posterior wings.

In certain of the Rhynchota (Heteroptera) the base of the fore wing is tough, leathery and opaque, whilst the tip is membranous: such wings are called *hemi-elytra*. In

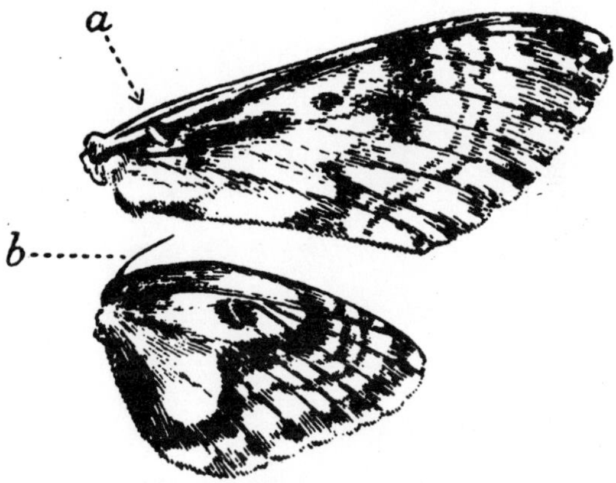

FIG. 12.—Wings of Death's Head Moth, showing coupling apparatus.

a. retinaculum : *b*. frenulum. British Museum Guide.

the Earwigs and other Orthoptera the fore wings are completely leathery, but not hard and chitinized like elytra: they are known as *tegmina*.

The posterior pair of wings in the true Flies or Diptera take the form of short slender structures, the *halteres* or *balancers*.

A number of species of insect are *apterous*—that is, they possess no wings at all in the adult state. This wingless condition may be due to two causes. Thysanura (Bristletails) and Collembola (Springtails) are supposed to have broken away from the main developing body of insects before wings were evolved. Bristletails, therefore, never at any stage of their history possessed wings. Their wingless condition is consequently said to be *primitive*.

There is a good reason, however, for assuming that fleas and lice originally possessed wings, but that through degradation these appendages have been lost. The wingless condition of fleas and lice is therefore acquired, and is said to be *secondary*.

CHAPTER IV

THE MOUTHPARTS

BOTH from the point of view of the systematic entomologist and also the practical farmer, a knowledge of the mouthparts of insects is of considerable importance. The mouthparts of different insects vary very considerably, and these differences are made use of in classifying the various species. The construction of the mouthparts also determines the nature of the damage which insect pests cause to farm or garden crops, and, as will be shown later, it has an important bearing upon control measures directed against injurious insects.

The mouthparts of insects can be classified in accordance with their functions. According to this scheme there are three more or less distinct groups, viz. :—(1) Those intended for biting purposes, (2) those modified for sucking purposes only, and (3) those adapted for piercing and sucking.

Those mouthparts included in group (1) all agree in one respect—the relatively great size of the jaws or mandibles (see Figs. 13 and 14). For this reason mouthparts adapted for biting are called *mandibulate.* Fig. 13 represents the construction of the mouthparts of the Queen Wasp, and Fig. 14 those of the Cockroach : the general scheme is the same in both insects, but differs in detail.

In both cases the parts are more or less completely enclosed between an upper lip (*labrum*) and a lower lip (*labium*). Immediately under the labrum are the two jaws or *mandibles,* consisting of hard, toothed plates of chitin. These jaws are controlled by powerful muscles, but they are capable of movement in one plane only. They can be moved outwards and inwards, but not in an up-and-down direction. Underneath the mandibles comes a pair of structures called the *first maxillæ.* Each maxilla is attached to the head by the *cardo* (*C.*), which acts as a hinge. To this is attached the stalk or *stipes* (*St.*), which

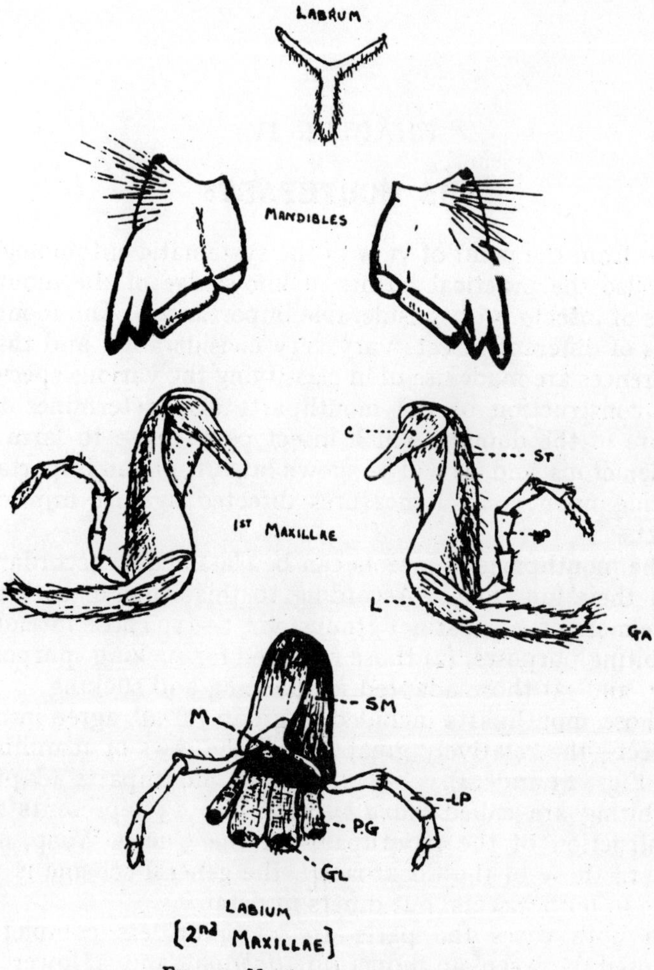

FIG. 13.—Mouthparts of a Queen Wasp.
C. Cardo : *St.* Stipes : *L.* Lacinia : *MP.* Maxillary palp : *Ga.* Galea : *SM.* Sub-mentum : *M.* Mentum : *LP.* Labial palp : *Pg.* Paraglossa :, *Gl.* Glossa.

bears three structures. Towards the outer edge arises the *maxillary palp* (*MP.*), composed of from one to seven joints. In thc wasp each palp has six joints, in the cockroach five joints. The maxillary palps are sensory organs. At the distal end of the stipes arise two structures. The outer one

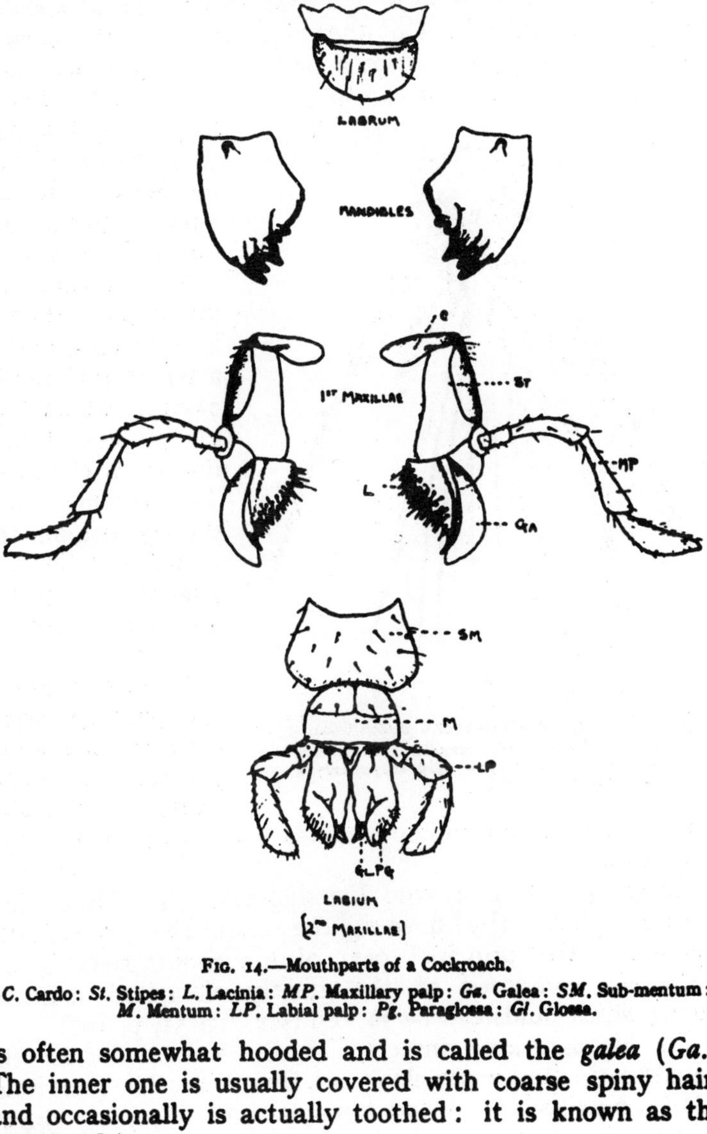

FIG. 14.—Mouthparts of a Cockroach.

C. Cardo: *St.* Stipes: *L.* Lacinia: *MP.* Maxillary palp: *Ga.* Galea: *SM.* Sub-mentum: *M.* Mentum: *LP.* Labial palp: *Pg.* Paraglossa: *Gl.* Glossa.

is often somewhat hooded and is called the *galea* (*Ga.*). The inner one is usually covered with coarse spiny hairs and occasionally is actually toothed: it is known as the *lacinia* (*L.*).

The labium consists of a pair of *second maxillæ* which have become fused together medianly. The joint of attachment takes the form of a broad plate, the *sub mentum* (*SM.*). To this is fixed another plate known as the *mentum* (*M.*). The mentum carries a pair of sensory *labial palps* (*LP.*), comprising from one to three joints, and the *ligula*, made up of a pair of *paraglossæ* (*Pg.*) on the outside and a pair of *glossæ* (*Gl.*) on the inner side.

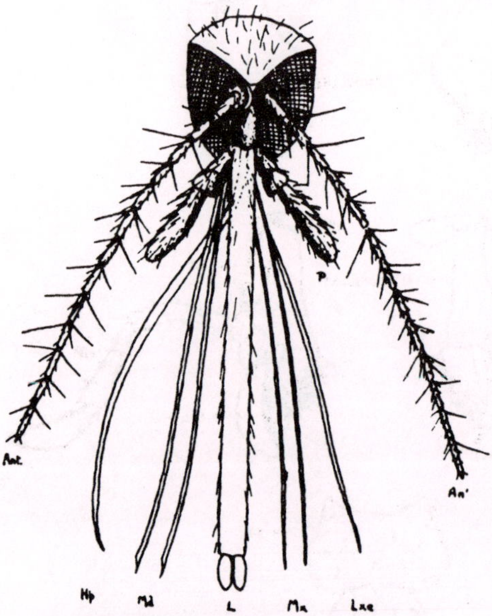

FIG. 15.—Mouthparts of Common Gnat.
Md. Mandibles: *Mx.* Maxillæ: *L.* Labium: *Lxe.* Labrum-epipharynx: *Hp.* Hypopharynx: *Ant.* Antennæ: *P.* Palp.

The labrum and labium form the top and bottom boundaries of the mouth cavity, whilst the mandibles and maxillæ are at the sides. The region immediately behind the mouth is called the *pharynx*; it is the anterior part of the intestine.

From the floor of the pharynx arises a short tongue-like process, the *hypopharynx* or *lingua*. The duct from the salivary glands has its opening in this structure.

In insects such as the Lepidoptera, which live upon nectar secreted by flowers, the mandibles are greatly reduced. The principal part of the mouthparts is the *proboscis*, composed of the greatly elongated maxillæ, along which fluids can be sucked (see Fig. 51, p. 100).

In blood-sucking insects like the common gnat the mouthparts are greatly elongated (Fig. 15). The mandibles (*Md.*), maxillæ (*Mx.*) and hypopharynx (*Hp.*) are drawn out into long bristles, which can be carried enclosed in the tube formed by the grooved surfaces of the *labrum-epipharynx* (*Lxe.*) and labium (*L.*). The maxillæ and mandibles make the puncture before feeding can commence,

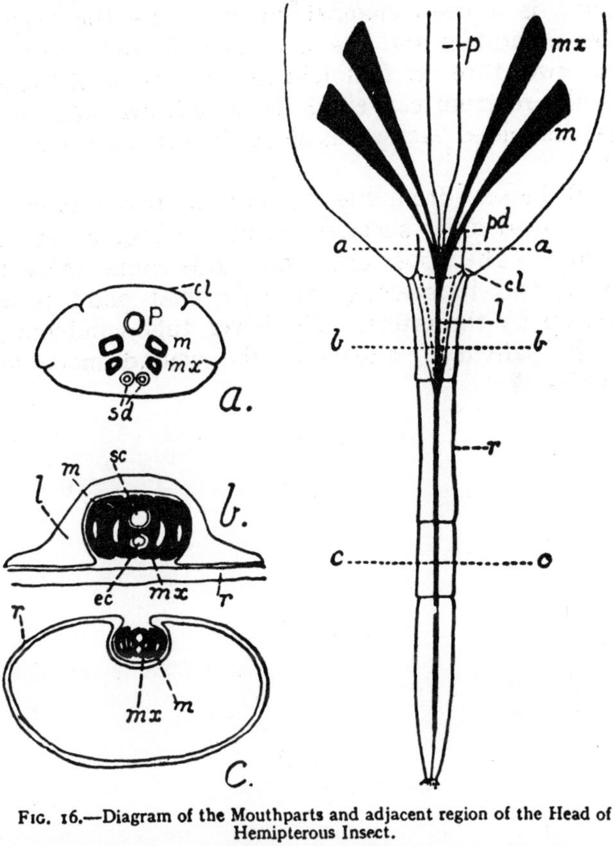

FIG. 16.—Diagram of the Mouthparts and adjacent region of the Head of a Hemipterous Insect.

On the left are transverse sections across the regions bearing corresponding lettering: the magnifications of these sections are not uniform. *cl.* clypeus : *ec.* ejection canal with salivary duct : *l.* labrum : *m.* mandible : *mr.* maxilla : *p.* pharynx : *pd.* pharyngeal duct : *r.* rostrum : *sd.* salivary ducts : *sc.* suction canal with pharyngeal duct.
From Imms, "General Text Book of Entomology."

whilst the *hypopharynx* conveys saliva to the wound. The hypopharynx and labrum, when opposed, form with their grooved surfaces a fine tube through which animal juices can be sucked up into the pharynx.

In the Rhynchota a further type of piercing and sucking

mouthparts is found, the complete apparatus being known as the *rostrum* (Fig. 16). The most conspicuous part is the labium, or lower lip, which is greatly elongated, and modified into an almost perfect cylinder. On the upper surface, however, is a deep channel, in which lie the very long, slender mandibles and maxillæ. The labium may consist of one, two, three or four joints. In certain of the Rhynchota the rostrum can either be withdrawn into the head or may be folded backwards along the ventral surface of the body.

It will be seen from Fig. 16 that the two maxillæ, when their grooved inner surfaces are opposed, give rise to very fine tubes. The upper or dorsal one is connected with the pharynx, and it is along this tube that plant juices are sucked up by the insect. The lower tube conducts saliva from the salivary glands into the wound made by the mandibles.

CHAPTER V

INTERNAL ANATOMY

INSECTS differ very greatly from the higher animals in two important respects. In the first place there is no internal arrangement of bones to give support and attachment to muscles and other tissues, but the rigid external covering or cuticle serves the same purpose. In places, notably the thorax and head, the integument may extend into the body cavity and become chitinized, thus acting as an endoskeleton, but, generally speaking, insects may be said to possess an exoskeleton only.

The second great departure from the architecture of the higher animals is to be found in the arrangement of the internal organs. In insects the main nervous system is ventrally situated, whilst the main vascular system is dorsal (Figs. 5 and 6); in the vertebrate animals this arrangement is reversed. Fig. 17 shows diagrammatically the relative positions of the chief organs in a typical winged insect. These organs include the Vascular or Blood System, the Digestive System, the Respiratory or Breathing System, the Nervous System and the Reproductive System.

Vascular System.—The internal economy of a typical insect can easily be studied by dissecting a cockroach. On cutting away the dorsal plates of the animal a comparatively wide tube of a darkish colour can be seen running from the head to the posterior part of the body. This is the great dorsal blood-vessel, which in the anterior portion of the body is known as the *aorta* (Fig. 17, *Ao.*). Towards the hinder end the tube forms a series of contractile dilations which collectively form the 'heart' of the insect. The posterior end is closed, whilst the anterior end is always open.

This dorsally situated organ is in most insects the only

structure which at all resembles the veins or arteries of the higher animals. The circulation of the blood is correspondingly simple, and takes place in the following manner. Between the chambers of the heart are a number of openings or *ostia*. When a chamber dilates these ostia open and blood from the body cavity rushes into the heart. When the chamber contracts, it automatically closes the ostia and forces the blood into the chamber next in front of it. Since a contraction always starts at the hindermost dilation,

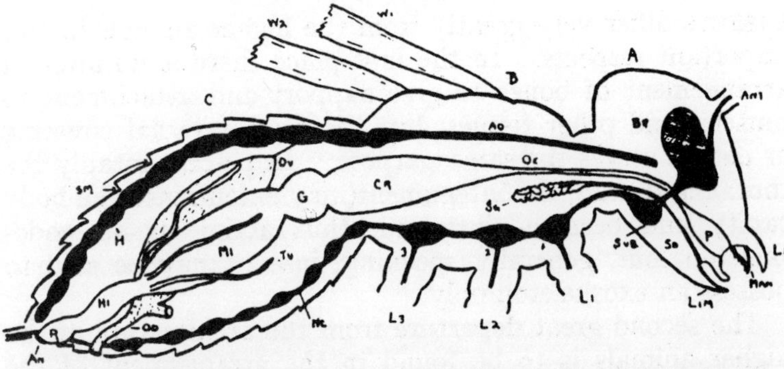

FIG. 17.—Diagram showing the Relative Positions of the Chief Organs of a Winged Insect.

A. Head: *B*. Thorax: *C*. Abdomen: *Ant*. Antenna: *Ao*. Aorta: *An*. Anus: *Br*. Brain: *Cr*. Crop: *G*. Gizzard: *H*. Heart: *HI*. Hind Intestine: *SM*. Intersegmental membrane: *Lim*. Labium: *Lr*. Labrum: L_1, L_2, L_3, Legs: *MI*. Mid Intestine: *Man*. Mandible: *NC*. Nerve Cord: *Ov*. Ovary: *Od*. Oviduct: *Oe*. Œsophagus: *P*. Pharynx: *R*. Rectum: *SG*. Salivary Gland: *SD*. Salivary Duct: *Sub*. Sub-œsophageal ganglion: *Tu*. Malpighian Tubules: 1, 2, 3, Thoracic ganglia.
Modified from a drawing by Imms.

the blood is propelled forward along the aorta and pours out at the open end of this vessel. It then finds its way back through the body cavity, bathing all the internal organs in its progress. At the same time blood is forced into the wings, legs and other extremities, but in these cases there are special tubes to convey and return the fluid. The blood carries digested food material to all parts of the body. It does not play a very important part in respiration, owing to the complicated series of air tubes which carry oxygen everywhere in the insect.

Digestive System.—If the heart be now carefully removed from the cockroach undergoing dissection, the whole of the interior of the body appears to be occupied by a mass of whitish tissue. This tissue is composed partly of the so-called 'fat-body' and partly of *tracheæ,* which will be described later.

The fat body, as its name implies, consists of fatty tissue. It is most fully developed in larvæ and in adult insects which hibernate. It serves as a store of nutriment, and

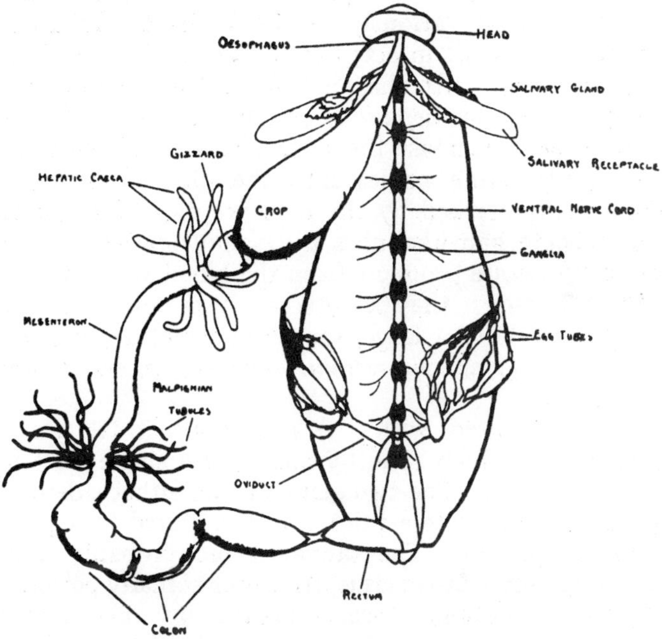

FIG. 18.—Cockroach dissected from the Dorsal Surface.

there is evidence to show that it plays an important part in excretion.

On cutting away the fat body the alimentary canal can be examined (Figs. 17 and 18). In essentials this consists of a more or less convoluted tube which connects *mouth* and *anus*. Immediately behind the mouth is a slight swelling, the *pharynx* (Fig. 17, *P.*). This passes into a straight tube, the *œsophagus,* which at its hinder end swells out into a thin-walled storage organ known as the *crop*. Behind this is a smaller structure, the *gizzard,* which has very thick, muscular walls, provided on the inside with a number of chitinous projections. According to some writers the gizzard acts in a crushing capacity, reducing hard food material to a fine condition. More recent research has shown that in many cases the gizzard acts rather as a strainer, only allowing food in a suitable condition to enter the stomach.

From the gizzard there proceeds, in the cockroach, a wide tube, known variously as the stomach, mid-gut, or *mesenteron.* In some insects the mid-gut is short, wide and sac-like (Fig. 17). At the anterior end are a number of hollow processes, the *hepatic* or *gastric cæca,* whose function appears to be to increase the surface area of the mid-gut.

The mid-intestine passes into the hind intestine, the junction being marked by the presence of a large number of long, delicate, hair-like tubes, the *malpighian tubes.* These tubes absorb waste products from the blood and pour them into the intestine: they are excretory organs. The hind intestine consists of a large, sacculate structure, the *colon,* and a smaller egg-shaped *rectum,* which opens, via the *anus,* to the exterior.

Closely associated with the alimentary tract are the *salivary glands,* which open by means of a single duct into the mouth cavity. The two glands lie on either side of the œsophagus and secrete a juice which is stored in a pair of salivary receptacles. The salivary fluid probably assists digestion: in some insects it is irritant or actually poisonous.

Respiratory System.—Insects breathe by means of air tubes which extend throughout the whole of the body and the appendages. These tubes, or *tracheæ,* communicate with the exterior through openings in the side of the body known as *spiracles*; as the tracheæ penetrate farther into the body they branch and rebranch to form very fine tubes or *tracheoles.* Each air-tube is lined with a thin layer of chitin, the *intima,* which is continuous with the chitin of the cuticle, and at each moult this tracheal lining is discarded. The tube is further strengthened and made elastic by a more or less spiral thickening which runs along inside the chitinous lining. The outer covering of the trachea consists of large, delicate, polygonal cells.

The spiracles, or external openings, occur in pairs, usually on the pleurites of the thorax and abdomen (Fig. 19). There are frequently two thoracic and eight abdominal pairs of spiracles, but the different orders of insects exhibit considerable differences in the numbers of spiracles they possess. The spiracle consists of the actual opening and the small chamber within, together with any arrangement of lips, valves and muscles which may be present to enable the spiracle to close. Some insects are able to close their spiracles, others are not. Each spiracle is surrounded by a ring of chitin, and in some cases the opening is protected by a number of delicately branched processes which project across it.

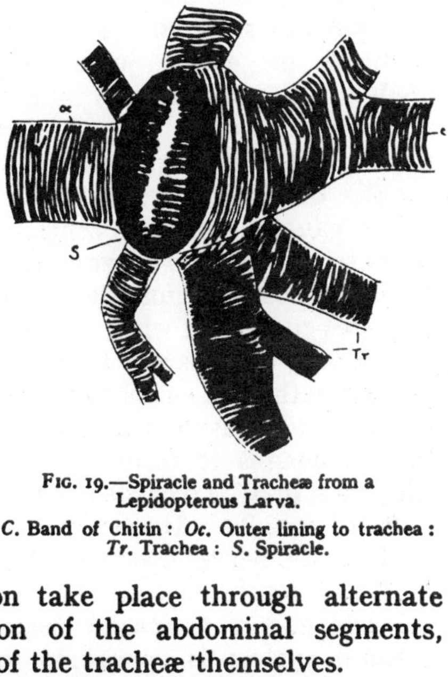

FIG. 19.—Spiracle and Tracheæ from a Lepidopterous Larva.

C. Band of Chitin : *Oc.* Outer lining to trachea : *Tr.* Trachea : *S.* Spiracle.

The tracheæ and tracheoles are intimately connected with the digestive apparatus, with the muscles, nerves, fat body, vascular and reproductive systems. They convey air to all parts of the body, supplying the oxygen required for the respiratory process and carrying away the carbon dioxide formed. Inspiration and expiration take place through alternate expansion and contraction of the abdominal segments, assisted by the elasticity of the tracheæ themselves.

Certain insects, notably the Springtails, have no spiracles or have them permanently closed. In such cases respiration is *cutaneous*, taking place through the thin integument over the whole of the body surface. In many aquatic insects, particularly in the larval stages, breathing takes place through special abdominal gills.

Three main types of respiratory system are commonly recognized:

1. *Holopneustic* type.—In this type all the spiracles are

open, and there are usually two pairs of thoracic and seven or eight pairs of abdominal spiracles.

2. *Hemipneustic* type.—Insects with this type of respiratory system have one or more pairs of spiracles permanently closed. It is the prevalent system amongst larvæ.

3. *Apneustic* type.—Spiracles are either wanting or all permanently closed in this type, respiration taking place through the surface of the body wall.

The Nervous System.—The central nervous system consists of a nerve cord extending from the head to the posterior extremity of the body. With the exception of the brain it is ventrally situated. The cord consists of a double row of swellings, or *ganglia*, joined together longitudinally by a double line of nerve fibres called *connectives* (Fig. 18). The anterior ganglion is the *sub-œsophageal ganglion*, the next three are the *thoracic ganglia*, followed by the *abdominal ganglia*, of which there may be as many as eight.

The ganglia are nerve centres, from which extend nerve fibres to muscles and other organs. Each ganglion controls, or is responsible for despatching and receiving nerve impulses concerned with, a definite area of the body.

From the front part of the sub-œsophageal ganglion, which lies beneath the œsophagus, arise two connective nerve fibres, which pass upwards around the salivary ducts and the œsophagus to form a connecting link with the brain. The brain may be regarded as the result of the fusion of three pairs of ganglia. It is dorsally situated in the head, and controls the nerves connected with the eyes and antennæ.

In addition to the central nervous system there is a *visceral* nervous and a *peripheral sensory* nervous system. The visceral system is connected with the brain and passes dorsally to the heart and to the fore gut. The peripheral sensory nervous system consists of nerve fibres and nerve cells lying in the integument of the body just below the *hypodermis*.

Reproductive System.—In insects the sexes are distinct. The female reproductive organs (Fig. 18) consist essentially of the *ovaries* and the ducts necessary to convey the eggs out of the body of the female. The ovaries are paired structures lying on either side of the posterior part of the alimentary canal. Each ovary consists of a number of egg tubes or *ovarioles*, which open into the *oviduct*. In the Cockroach there are eight egg tubes to each ovary. It is in the egg tubes that the eggs are formed, lying in a chain with the largest eggs nearest the oviduct. As they pass down the lower part of the duct the eggs are fertilized by *spermatozoa*, which at pairing time are deposited by the male in the *spermatheca*. This is a pouch connected with the duct, specially designed for storing spermatozoa, so that fertilization of the eggs may proceed over a long period. The oviduct opens to the exterior about the eighth or ninth abdominal segment.

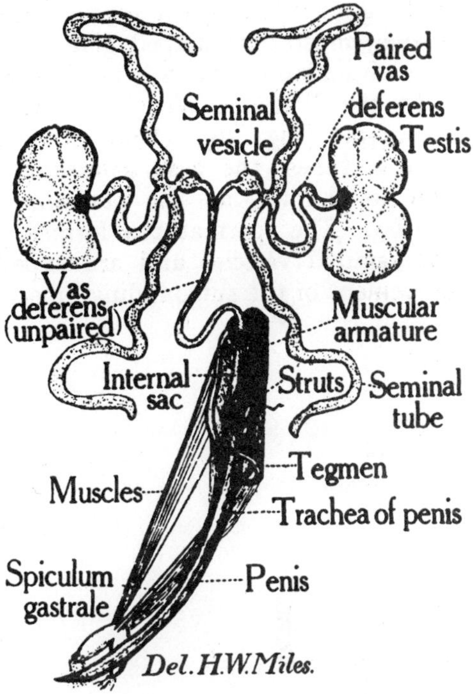

FIG. 20.—Reproductive organs of male Apple Blossom Weevil (*Anthonomus pomorum*).

Female insects deposit their eggs with considerable care, and in this operation many are assisted by an elaborate apparatus called the *ovipositor*. This consists of paired structures developed from the eighth and ninth abdominal segments. These take the form of long narrow valves, channelled on the inner surface, forming a tube down which the eggs can be passed. In the Wasp the ovipositor is not readily seen: in the Wood Wasp the ovipositor forms a conspicuous terminal appendage. In the Sawflies the valves of the ovipositor are modified into saws, by means of which slits can be cut in vegetable tissue: in these slits the eggs are frequently laid. In the Honey Bee the ovipositor of the ' worker ' females is modified into the well-known sting.

The male reproductive organs consist of the paired *testes*, the *ducts* and *accessory glands*. In the Cockroach the testes can only be discerned in young specimens: they take the form of long, narrow, lobed structures. The most striking structure of the reproductive organs of this insect is the ' mushroom-shaped gland,' or accessory gland, the exact function of which is not known.

In the Blossom Weevil the testes are kidney-shaped structures (Fig. 20). They are connected by a pair of narrow tubes, the *vasa deferentia*, which swell out into pouches known as the *vesiculæ seminales*, or *seminal vesicles*. The spermatozoa formed in the testes collect in the seminal vesicles and are expelled by an ejaculatory movement of the surrounding muscles via the *penis*.

CHAPTER VI

METAMORPHOSIS

THE majority of insects start life in the form of a fertilized egg or *ovum*, which is passed out of the body of the female and deposited on leaves, twigs or in the ground. In the case of an insect which parasitises an animal, eggs are laid in or upon the bodies of the unfortunate creature, which is said to be the *host* of the parasite.

After the hatching of the egg the developing individual undergoes a series of more or less complicated changes in size and structure before becoming adult, or sexually mature. To the series of changes which take place in the life history of an insect the term *metamorphosis* has been applied.

A study of the metamorphoses of insects is of great importance, since control measures cannot be adopted successfully against pests unless their life histories are known with a reasonable degree of accuracy.

In some insects the form which hatches out from the egg very closely resembles the adult insect, or *imago*. Metamorphosis in such cases is *slight*. In other insects very complicated changes are undergone during growth, and metamorphosis is consequently *complex*.

In most textbooks complex metamorphosis is called 'complete' and slight metamorphosis is called 'incomplete' metamorphosis. Since the series of changes which an insect undergoes during its life-cycle is complete, so far as that insect is concerned, whether the changes are simple or complicated, the terms slight and complex are to be preferred.

Slight Metamorphosis.—The Capsid Bug exhibits an easily observed slight metamorphosis (Fig. 21). The animal which emerges from the egg of a capsid bug very closely resembles its parents. The shape of the body, the structure of the legs, the whole general appearance of the young insect are similar to those displayed by the imago except that the newly hatched specimen does not possess wings or developed sex organs. A young insect of this nature is called a *nymph.*

The young bug feeds voraciously and grows so much that its cuticle can no longer enclose the internal organs. In order to allow an increase in size to take place, an *ecdysis* or *moult* becomes necessary, since the chitinized cuticle secreted by the true skin is incapable of expansion.

Before an ecdysis occurs, there is a preliminary separation of the true skin from the original cuticle. The space between these two layers becomes filled with a watery fluid. The skin, which during growth has become much folded in the limited space allotted to it, proceeds to form a new cuticle over its whole area. When this has been completed the old cuticle is split, and the enclosed insect crawls out of the now useless shell, called the *exuvia.* The folded skin expands, the chitin hardens, and a new stage or *instar* has been inaugurated. This instar differs slightly from the one preceding it, and after one or two more moults the wings appear as small buds. At successive moults these wing buds increase in size; the sex organs develop and the imaginal stage is reached.

An instar is therefore the form assumed by an insect during the intervals between successive ecdyses. The first instar occurs between the time of hatching from the egg and the first ecdysis. The imago is the final instar.

During ecdysis the chitinous linings to the fore gut, the hind gut, and the tracheæ (p. 24) are sloughed off as well as the cuticle. An important point in the metamorphosis of this insect is that the wings become externally visible at an early stage in the life history.

Insects which share this peculiarity are grouped together into the division *Exopterygota,* i.e. insects with external wing development.

Complex Metamorphosis.—The life history of the common Cabbage White Butterfly displays a very different state of affairs (Fig. 22). From the developing egg comes a *cater-*

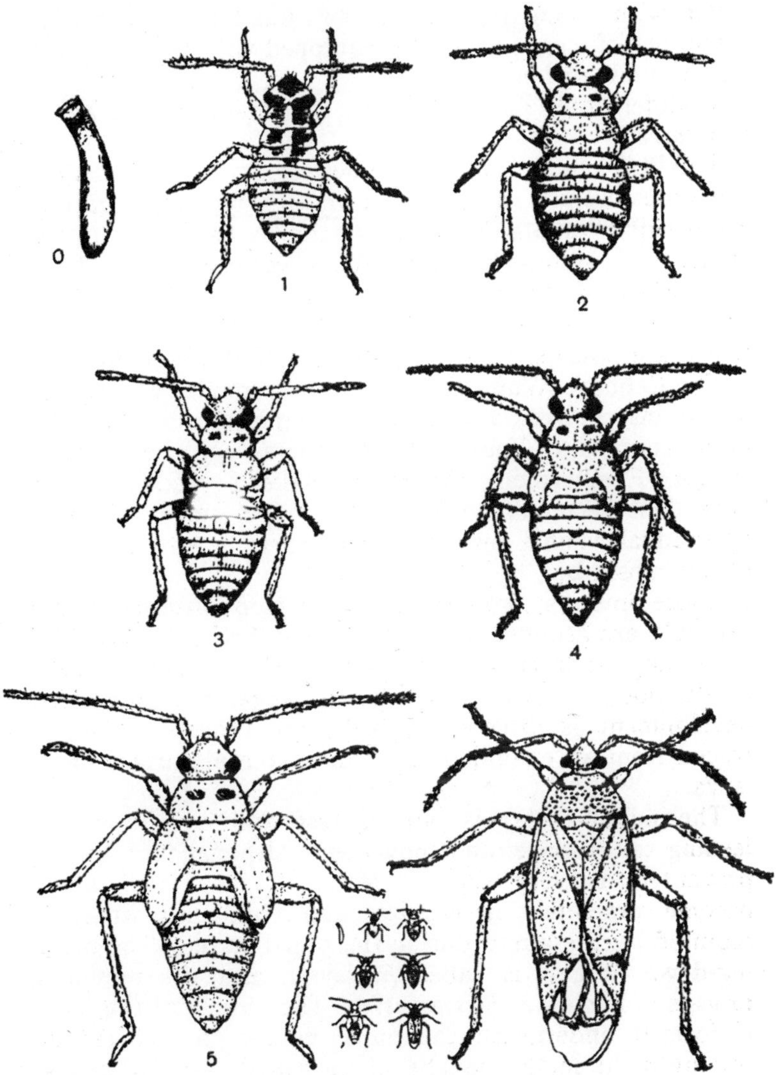

FIG. 21.—Metamorphosis of a Capsid Bug (*Plesiocoris rugicollis*).

O. Egg of Orthotylus marginalis. 1-5. Instars: 6. Adult of *Plesiocoris*. Inset. natural size. After Petherbridge.

pillar, with a long, segmented, wormlike body, which bears a number of legs, and a head equipped with powerful jaws and a number of ocelli.

It differs fundamentally from the imago and is known as a *larva*.

This caterpillar is a ravenous feeder and undergoes numerous moults necessitated by its continual growth. Eventually, at its last moult, the larva turns into a short, thick, spindle-like body, the *chrysalis* or *pupa*. Upon the surface of this pupa can be made out the outlines of the wings, legs, mouthparts, abdomen, and other parts which are possessed by the imago which develops from it.

The Cabbage White Butterfly thus provides an example of an insect with a complex metamorphosis. The life-cycle consists of four distinct phases—*egg*, *larva*, *pupa*, *imago*. Furthermore, the wings make a sudden appearance, first becoming visible in the pupal stage. There is no gradual, visible development as there is with the wings of the capsid bug.

Those insects whose wings develop internally and invisibly are grouped into the division *Endopterygota*.

In the Endopterygota wings and other structures begin to develop actually in the larval stage, but since their development is directed inwards, the process cannot be traced from an examination of the external surface of the larva.

The pupal instar is one of rest and transition. No feeding takes place, and movement, although not entirely prevented, is very slight. Many of the larval tissues become completely disorganized and rebuilt into structures required by the adult, during the pupal state. The breaking-down process is called *histolysis*, and the rebuilding process is known as *histogenesis*. The cuticle of the imago is formed underneath the pupal cuticle, but it remains soft until the imago sloughs off the pupal case and exposes itself to the air.

In certain orders of insects are to be found species the females of which are capable of bringing forth living young. Such females are said to be *viviparous* as opposed to the more usual type of *oviparous* or egg-laying females. In

Fig. 22. - Metamorphosis of cabbage white Butterfly, showing eggs (highly magnified), larva, pupa, male and female adults together with a pupa surrounded by the cocoons of a hymenopterous parasite (*Microgaster glomeratus*).

viviparous females the eggs are retained until the embryo has developed into the nymph: oviparous females pass out the newly formed eggs from their bodies, so that embryonic development takes place outside the parent.

Parthenogenesis.—Insects frequently exhibit the phenomenon of *parthenogenesis*, or reproduction without fertilization (Aphides, etc.). The normal reproduction of insects, as of most animals, involves the fusion of the male reproductive cell or sperm with the female reproductive cell or ovum. In some insects, however, the ovum develops without the stimulus of the male sperm.

In some families males are quite unknown: in others they are infrequent and only occur at certain times of the year. Parthenogenesis may be the normal method of reproduction, it may be sporadic, or it may alternate with the more normal sexual reproduction.

In some insects such as species of Hymenoptera males are produced from unfertilized, and females from fertilized eggs: in others both males and females can be produced parthenogenetically, as in certain Tenthredinidæ. In the Aphididæ viviparous females reproduce parthenogenetically.

Pædogenesis is the production of young by immature individuals. Some pupæ and larvæ have been observed to be capable of pædogenetic reproduction, either by the production of further larvæ or of eggs.

The Egg.—The eggs of insects vary enormously in size and shape, colour and markings. The majority are roughly spherical: others may be cylindrical, like those of the Mangold Fly. Some are laid singly (Blossom Weevil), others in clusters (Lackey Moth, Cabbage Butterfly) and some at the end of long stalks (Lace-wings): many are firmly glued to hair or feathers (Lice, Warble Fly, etc.).

The Larva.—Larvæ of insects take many different shapes, and special terms have been used to distinguish them. The true *caterpillar* is the larva of one of the Lepidoptera. It consists usually of thirteen segments (exclusive of the head), but since the last segment but one is frequently very difficult to see, the body of many caterpillars appears to consist of the head and only twelve body segments. The first three segments, which correspond in some ways to the three thoracic segments of the imago, each bear a single pair of jointed legs. The last, or *anal*, segment bears one, and the middle segments of the body bear not more than four pairs of thick, fleshy, unjointed suckers, or *pro-legs* (see Fig. 23, *A*). The fifth segment never bears pro-legs.

In some families of Lepidoptera the pro-legs in the centre of the body are reduced to a single pair, as in the 'loopers' of the Winter Moth (Geometridæ) (Fig. 23, *B*). The caterpillar of the Y Moth has two pairs of central pro-legs.

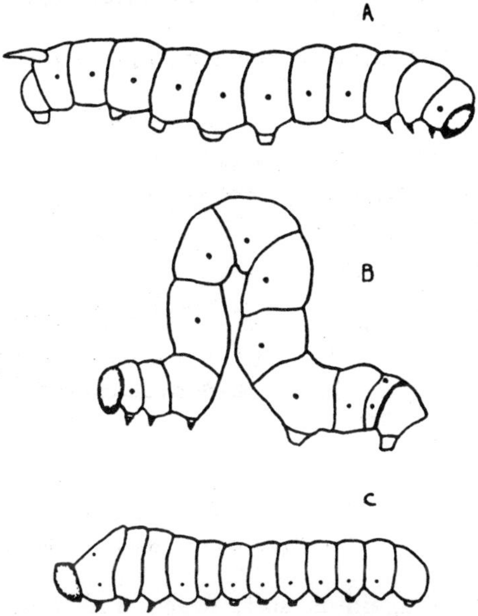

FIG. 23.—True and False Caterpillars.

A. Typical Lepidopterous larva. *B*. 'Looper' larva of Geometridæ. *C*. 'False Caterpillar' of a Sawfly.

Ignoring the head segment and assuming the presence of only twelve body segments, we can construct a 'pedal formula'[1] to enable us to recognize these larvæ. The typical formula for Lepidoptera is *3004001*. The first number refers to the three pairs of true, jointed legs. The next two cyphers show that the following two segments bear no pro-legs. Then come four segments each bearing pro-legs; two segments without pro-legs, and an anal segment with a pro-leg. The formula for the larva of the Geometridæ is *3000001001*.

The larvæ of some Saw-flies (Tenthredinidæ) closely resemble true caterpillars. They have, however, more than four pairs of central pro-legs, their formula being *30601* or *308*. They are known as *false caterpillars* (Fig. 23, *C*).

[1] Devised by Professor F. J. Cole, D.Sc., F.R.S., of the University of Reading.

Grubs are the larvæ of certain Hymenopterous and Coleopterous insects. They are usually short, fat, soft, fleshy, saccate, legless and of a dirty white colour (Fig. 24, *A* and *B*). The name is also applied to the larvæ of the Chafer tribe, which though very sluggish, yet possess three pairs of conspicuous legs. The term grub is not a precise one: it is frequently applied to any soil-living larva. The larva of the Crane Fly is often referred to as 'The Grub.'

The larva of the true flies takes the form of a *maggot.*

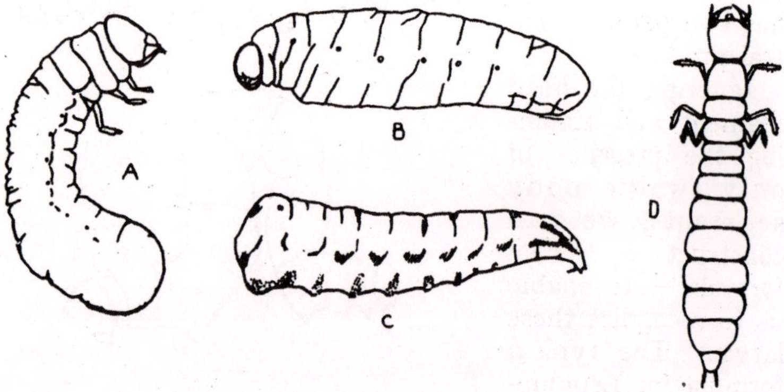

FIG. 24.—Types of Larvæ.
A. Cockchafer. *B*. Wasp. *C*. Fly. *D*. Rove Beetle.

This again is soft, fleshy and whitish, but it is usually much longer than broad, and tapers to a point at the head. The head may or may not be distinct. No legs are present (Fig. 24, *C*).

Larvæ are provided usually with mandibulate mouth-parts, but in larvæ such as the grubs of social Hymenoptera, which are fed upon more or less liquid food, the mandibles are very much reduced or even actually wanting. Larvæ are also provided with antennæ, which reach their greatest development in those types which are predaceous, such as Staphylinid larvæ (Fig. 24, *D*). Those larvæ which do not move about a great deal have very short antennæ.

The Pupa.—Different types of insects have different kinds of pupæ (Fig. 25). In some, antennæ, legs, etc., are not firmly stuck down to the body: such pupæ are said to be *free, libera,* or *exarate.* They are common in Coleoptera, Hymenoptera and other orders. In Lepidoptera, particularly, the appendages of the pupa are stuck firmly down to the body, thus forming an *obtect* pupa.

The pupa of the Butterflies takes the form of a somewhat angular, pale, often shiny structure, the *chrysalis.* Other insects either spin a covering of silk round the pupa or

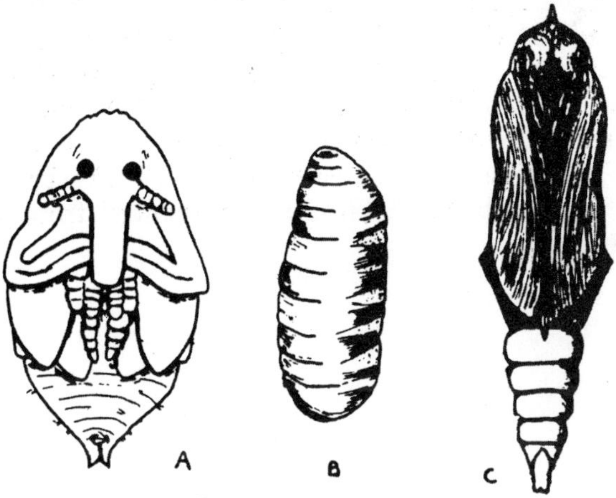

FIG. 25.—Types of Pupæ.
A. Free pupa of Apple Blossom Weevil. *B.* Puparium of Diptera. *C.* Obtect pupa of Lepidoptera.

surround it with a covering of soil particles plastered together with an excreted fluid, thus forming *cocoons.* Some true flies pupate in the old larval skin, forming a ***puparium*** or *coarctate pupa.*

The length of time that an insect spends in any one stage varies enormously. In some cases the complete life-cycle is completed two or three times in the course of a single year. Others require several years to complete one cycle. The egg of the House-fly develops in a few hours: in some of the Lepidoptera eggs do not hatch out in less than six months. Some larvæ, such as the Wireworm, may live three or more years in the soil before pupating.

Hibernation.—During the winter months most insects are inactive. Others, such as the Winter Moth and

Mottled Umber Moth, do not emerge until cold weather has set in. Insects may over-winter in any of the four stages of metamorphosis. The Flea Beetles, the Apple Blossom Weevil, etc., over-winter as adults; the majority of moths and some beetles as pupæ; the Gooseberry Moth and the Chafer as larvæ, whilst others pass the cold weather as eggs (Lackey Moth, Capsid Bugs, etc.).

Cold has very little effect upon larvæ and pupæ. Leather-jackets have been artificially frozen solid, so that they chinked and shattered into fragments when dropped upon a sheet of glass: nevertheless upon being thawed out very gradually, a number of them came to life and behaved in a normal fashion. The old belief that a hard winter kills off insect pests is not well founded. Frosty weather favours insects, because birds are prevented from digging in search of them.

CHAPTER VII

CLASSIFICATION

In order to understand the relationship of insects to the rest of the animal kingdom, it will be necessary to give an outline of the way in which animals are classified.

The bodies of all living things are composed of *cells.* Some animals are exceedingly simple in structure and consist of a single cell only. Such unicellular animals are all grouped together into the sub-kingdom of the **Protozoa,** of which *Amœba* and *Monocystis* are examples.

All animals whose bodies are made up of more than one cell form members of the second sub-kingdom or **Metazoa.**

The Metazoa can be divided into two more or less distinct divisions. The first is built up of those animals which do not possess a backbone or internal bony skeleton, but whose bodies are either soft or provided with a chitinous exoskeleton. Such animals comprise the group **Achordata,** or more familiarly, **Invertebrata.** The second division of the Metazoa includes those animals which have a backbone and an internal bony endoskeleton: it is called **Chordata,** or **Vertebrata.**

Both the Invertebrates and the Vertebrates are built up of a number of groups of animals of distinct types: each group is called a *phylum.* The Invertebrates (Achordata) comprise the following phyla:—

1. Spongidæ—Sponges.
2. Cœlenterata—Jelly-fish, Corals.
3. Echinodermata—Starfish, Sea-urchins.
4. Vermes—Worms.
5. Mollusca—Shells, Snails, Slugs, etc.
6. Arthropoda—Crabs, Spiders, Mites, Insects, etc.

Vertebrate animals belong to the phylum Chordata, and are distributed amongst the following classes :—

1. Pisces—Fishes.
2. Amphibia—Toads, Frogs, Salamanders.
3. Reptilia—Snakes, Crocodiles, and Lizards.
4. Aves—Birds.
5. Mammalia—Mammals (including man).

The following scheme, which, of course, is of the very sketchiest, gives a general idea of the classification of the animal kingdom.

ANIMAL KINGDOM

- PROTOZOA
- METAZOA
  - *Invertebrates*
    1. Spongidæ
    2. Cœlenterata
    3. Echinodermata
    4. Vermes
    5. Mollusca
    6. Arthropoda
  - *Vertebrates*
    1. Pisces
    2. Amphibia
    3. Reptilia
    4. Aves
    5. Mammalia

Insects are included in the phylum **Arthropoda,** whose members share the following characteristics. The body is segmented and covered externally with chitin. Many of the segments carry paired appendages which are composed of a varying number of joints. The possession of these jointed appendages is, in fact, the chief feature of the phylum, so that the Arthropoda are known as the 'jointed limbed' animals. In addition, the vascular system is mainly dorsal, whilst the central nervous system consists of a simple form of brain and a ventral nerve cord (Figs. 5, 6, and 17).

The phylum Arthropoda comprises the following four classes :—

1. Crustacea—Crabs, Crayfish, Woodlice.
2. Arachnida—Spiders, Mites, Ticks.
3. Myriapoda—Centipedes and Millipedes.
4. Hexapoda or Insecta—Insects.

These classes may be distinguished by the following characteristics :—

1. The Crustacea are chiefly aquatic. They have at least five pairs of legs and two pairs of antennæ. Respiration is by means of gills.

2. The Arachnida have their bodies divided at most into two parts—*cephalo-thorax* (p. 273) and abdomen—whilst the adult has four pairs of legs. There are no antennæ.

3. The Myriapoda possess only a single pair of antennæ, but the body is worm-like and not differentiated in any way. Each segment bears one or two pairs of legs.

4. Insects are distinguished from the other three classes by the division of the body into three parts, Head, Thorax, and Abdomen, by the possession of six legs in the adult, and, usually, by the presence of wings.

About 450,000 different kinds of insects have been described and named. Those individuals which resemble each other so closely that they form a natural division are said to constitute a *species*. Species, in their turn, form natural groups, called *genera* (sing. *genus*), and these in a similar fashion constitute *Families*, *Sub-orders*, and *Orders*.

The name given to a Family of insects usually terminates in *-æ*, whilst that of an Order or Sub-order ends in *-a*. The Turnip Flea Beetle referred to below belongs to the Family *Chrysomelidæ* of the Order *Coleoptera*.

Every known insect receives two scientific names, usually in Latin or Greek, so that it can be understood and recognized in all civilized countries. Thus the common Turnip Flea Beetle is known scientifically as *Phyllotreta nemorum*. The first word forms the *generic* and the second the *specific* name. Unfortunately different names have been and are still being given to the same individual by different entomologists. This has led to much confusion, and in doubtful cases both generic names are sometimes given. For example, the Turnip Flea Beetle may be referred to as *Phyllotreta* (*Haltica*) *nemorum*, the name in brackets referring to the genus in which the animal was at one time included.

Numerous attempts have been made from time to time to classify insects and reduce them to some kind of order. In any scheme of classification special attention must be paid to the presence or absence of wings, the nature of the mouthparts, and the complexity or otherwise of metamorphosis. There can be no hard and fast scheme of classification, as fresh facts are continually coming to light which necessitate the revision of original estimates. In most cases, the finer points of classification nearly all depend upon the personal predilections of the author.

The following classification, which recognizes twenty-three different orders of insects, is taken, with slight modifications, from a "General Text Book of Entomology," by A. D. Imms. The British orders of economic importance are printed in heavier type.

Sub-Class I.—APTERYGOTA

Apterous insects, the wingless condition being primitive, with slight or no metamorphosis.

Order 1. Thysanoptera—Bristletails.
,, 2. Protura.
,, 3. **Collembola**—Springtails.

Sub-Class II.—PTERYGOTA

Winged insects which are sometimes secondarily apterous (see p. 14). Metamorphosis very varied, rarely slight or wanting.

Division I.—Exopterygota

Insects passing through a simple and sometimes slight metamorphosis, very rarely accompanied by a pupal instar. The wings develop externally and the larvæ are called nymphs (see p. 30).

Order 4. Orthoptera—Cockroaches, Grasshoppers, etc.
,, 5. Dermaptera—Earwigs.

Order 6. Plecoptera—Stone-flies.
„ 7. Isoptera—White 'Ants' or Termites.
„ 8. Embioptera.
„ 9. Psocoptera—Book-lice.
„ 10. **Anopleura**—Biting and Sucking Lice.
„ 11. Ephemeroptera—May Flies.
„ 12. Odonata—Dragon Flies.
„ 13. **Thysanoptera**—Thrips.
„ 14. **Rhynchota** (Hemiptera)—Plant Bugs.

Division II.—ENDOPTERYGOTA

Insects passing through a complex metamorphosis always accompanied by a pupal instar. The wings develop internally and the larvæ are usually specialized.

Order 15. **Neuroptera**—Lace-wings.
„ 16. Mecoptera—Scorpion Flies.
„ 17. Tricoptera—Caddis Flies.
„ 18. **Lepidoptera**—Butterflies and Moths.
„ 19. **Coleoptera**—Beetles.
„ 20. Strepsiptera—'Stylops.'
„ 21. **Hymenoptera**—Ants, Bees, Wasps, etc.
„ 22. **Diptera**—Two-winged Flies.
„ 23. **Aphaniptera**—Fleas.

Part II

ECONOMIC SECTION

CHAPTER VIII

COLLEMBOLA—SPRINGTAILS

SPRINGTAILS are tiny, wingless insects usually less than 5 mm. in length. They differ from all other insects in that the abdomen consists of six segments only, and since the first two, three or four abdominal segments in some species are fused together, the abdomen of certain springtails appears to have as few as three segments.

Compound eyes are absent, but a group of ocelli, each one comprising eight or less, is situated on either side of the head. The antennæ consist typically of four joints. The body is usually covered with hairs, sometimes with scales. Tracheæ are frequently absent, and respiration is largely cutaneous.

The Mouthparts are mandibulate and are capable of being withdrawn into the head.

Wings are never present in Collembola, and this apterous condition is reckoned to be primitive. In this respect springtails differ from the fleas and wingless forms of lice, whose apterous condition is considered to be secondary (see p. 14).

Metamorphosis.—There is no metamorphosis of any sort.

Locomotion.—The legs of the springtail are frequently feebly developed compared with the size of the body, and locomotion takes place by a special 'springing organ' carried on the underside of the body. This takes the form of a *furcula*, which arises from the distal end of the fourth segment. It consists of a basal portion bearing two arms, which normally lie forward along the ventral surface of the fourth and third segments, being held in position by a trigger-like hair or *hamula* situated on the third segment (Fig. 26). When the furcula is released from the hamula, muscular contraction causes the furcula to strike downwards and backwards with considerable violence, with the result that the insect is propelled several inches into the air. The name 'Springtail' is due to this peculiarity, but all members of the order are not capable of progression by springing.

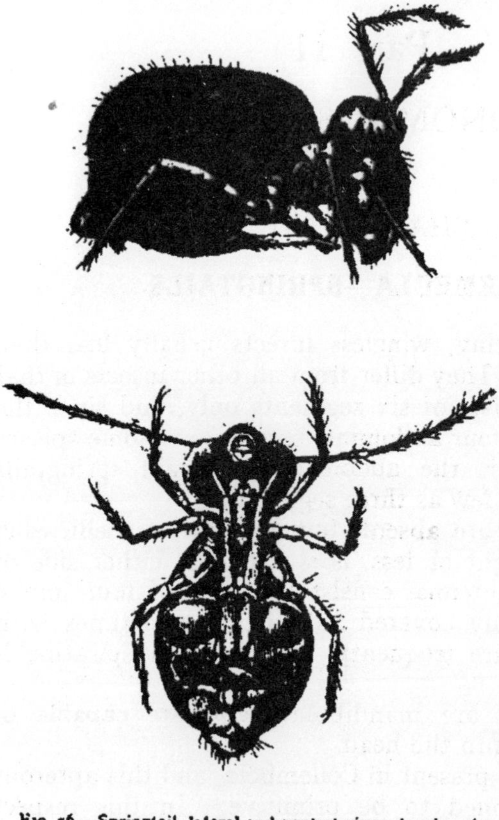

FIG. 26.—Springtail, lateral and ventral views showing the jumping organ.
British Museum Guide.

On the underside of the first abdominal segment in all Collembola is a small, bilobed structure called the *ventral tube*. This can be dilated at will, and it probably enables the insect to cling to smooth surfaces.

Plants attacked and damage done.—Springtails are known to attack many farm crops, but the amount of damage they cause is, as a rule, not sufficiently serious to warrant the employment of special counter measures. Recently, however, springtail attack upon mangolds and sugar beet has become very serious in several parts of the country, the most destructive species being individuals belonging to the genera *Bourletiella* and *Sminthurus*. The insects attack

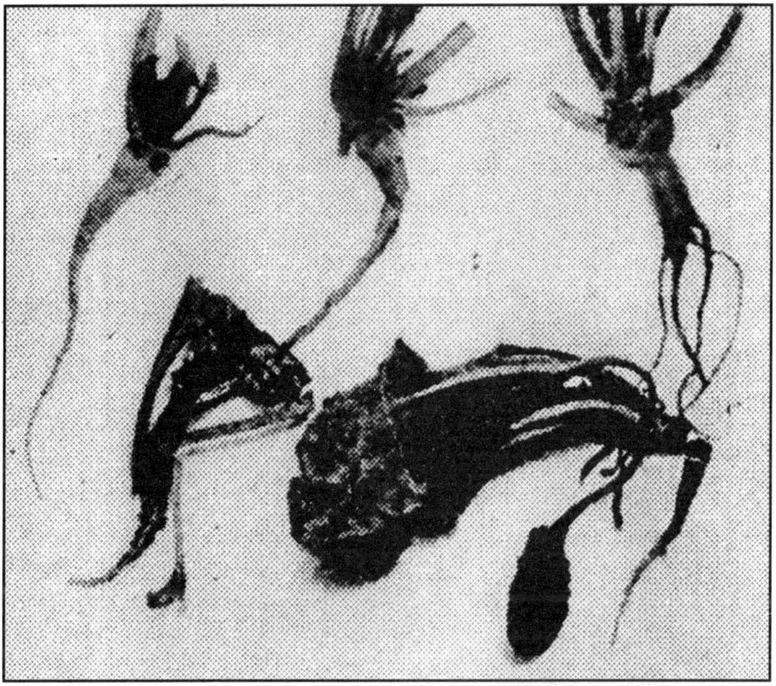

FIG. 27. - Mangolds attacks by Springtails. Reproduced by permission of H.M. Stationary Office and the Ministry of Agriculture and Fisheries.

both the leaves and the root of mangolds and sugar beet during the seedling stage. The leaves are eaten away in patches, but the reduction in leaf area is usually not sufficient to kill the seedling, although it may hinder its growth. The chief damage is caused to the root, at or slightly above ground level. Wherever springtails have been feeding a

FIG. 28. Improved machines for trapping springtails. (W. Maldwyn Davies.)

constriction appears, due to the chewing away of the epidermal tissues, and the subsequent enlargement of the root above and below the point of damage (Fig. 27). The portion below the constriction goes black, and as the seedling is knocked about by wind or during singling, the root snaps off and the seedling dies.

These characteristic symptoms must not be confused with those following an attack by the Pigmy Beetle (*Atomaria linearis*). In the latter case there is no constriction, but the whole of the root below the point of attack turns black and spindly (Fig. 74B).

Control Measures.—On a small scale springtails can be trapped in large numbers by employing devices similar to those shown in Fig. 28. That on the left consists of three boards built up into an arch which will just clear an ordinary mangold drill. A wire is stretched across the front end in such a way as to disturb the seedlings over which it passes, thus causing the insects to jump. The inside of the arch is kept covered with wet tar which traps the insects. The other apparatus explains itself: the sacks must be covered with tar in this case. Both kinds of apparatus may also be used successfully against the Turnip Flea Beetle (see p. 155).

Mangolds may be rendered obnoxious to springtails and other insects by dragging over them sacks soaked in paraffin oil.

There is a certain amount of evidence to show that sugar beet drilled upon ridges suffers more damage from springtail attack than that sown on the flat. An attacked plant becomes very brittle at the constricted part, and if during singling the free part of the seedling falls an inch or so from the top of the ridge, a breakage at the pinched area is very prone to occur.

Those species of *Sminthurus* which attack currants and apples may be trapped in large numbers underneath scooped-out orange rinds.

CHAPTER IX

ANOPLEURA—BITING AND SUCKING LICE

INTO this single order have been grouped the Biting Lice and the Sucking Lice, which by many writers are considered to constitute separate orders.

Although the order can be split into two quite natural divisions, because of the differences in mouth structure, yet these two forms of lice have much in common. Both forms are apterous and devoid of ocelli. Their bodies are flattened, with the thoracic segments more or less fused together. Both forms are *ectoparasites*—they live in the feathers or on the skin of birds, or on the skin of mammals.

They may be distinguished very readily by the size of the head. That of the Mallophaga is relatively broad and enormous, whilst the Siphunculata have small narrow heads.

The two divisions of the Anopleura are:

SUB-ORDER.—1. *Mallophaga* or Biting Lice.—Head large and crescent-shaped: mouthparts mandibulate: thoracic spiracles ventral: tarsus with one or two joints.

SUB-ORDER.—2. *Siphunculata* or Sucking Lice.—Head longer than broad: mouthparts modified for piercing and sucking: thoracic spiracles dorsal: tarsus always one-jointed.

SUB-ORDER.—1. MALLOPHAGA.—Bird Lice or Biting Lice.

These degenerate insects are usually less than 5 mm. long. Their eyes are much reduced. The prothorax is free; the meso- and metathorax partially fused. The tarsus has either one or two joints, and bears either one or two claws. The antennæ carry three, four or five joints.

Mouthparts are mandibulate, with large, toothed mandibles and simple maxillæ inserted on the ventral surface of the large head.

Wings *are never present, the apterous condition being* secondary.

Metamorphosis.—The life histories of the common kinds of Biting Lice are very similar, and metamorphosis is always simple. The insect spends the whole of its life upon the animal it parasitises. Small, rather longish eggs ('nits') are laid separately, each one stuck to a feather. From the eggs hatch out in from six to ten days nymphs, which very closely resemble the adults in shape and methods of feeding. After several moults they become fully grown, the process occupying two, three or more weeks.

The Mallophaga chiefly infest birds, but species belonging to two genera are known to occur on mammals There are about *1700 species* of Mallophaga, but not one parasitises man.

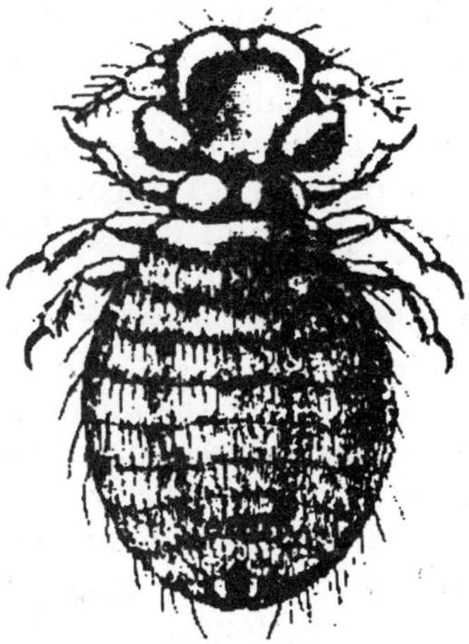

FIG. 29.—Biting Louse (*Trichodectes latus*) of the Dog. British Museum Guide.

The genus *Trichodectes* includes a number of individuals which attack domestic animals and other mammals. One species, *Trichodectes latus* (Fig. 29) occurs on the Dog, and may serve as a secondary host for one of the numerous Tape-worms which infest this domestic animal. Its antennæ are three-jointed, and the tarsus bears a single claw. Other species are found on the Horse, Ox and Sheep. They cause hair to fall off in large patches, but usually do not do so much damage as the Sucking Lice.

Eight distinct species attack poultry The most troublesome is *Menopon pallidum*, distinguished both on account of its active habits and its four-jointed antennæ, which are short and concealed beneath the head. The head is large and somewhat crescent-shaped. Its body (Fig. 30) is about

2 mm. long, with a broad, oval abdomen. Each segment is pale yellow in colour, with two bright fawn spots on either side. The abdomen bears numerous bristles. *Menopon* is the most active of the bird-lice and frequently crawls on to the hands of persons engaged in plucking or handling infested birds. It may cause irritation in this way for a short time, but it soon dies if removed from the fowl: it lives largely upon the quill epidermis of feathers.

The other species of bird-lice have five-jointed antennæ which project beyond the head. *Lipeurus variabilis* (Fig. 30) varies from 1·9 mm. in length, the female being larger than the male. It is the most slender of the common forms, with an abdomen less than half as broad as that of *Menopon*, and tapering to a point. In colour it is yellowish, with fawn spots and darker bands. This species lives between the barbs of the wing and tail feathers.

Goniocotes hologaster (Fig. 30) varies in length from just under to just over 1 mm. The head is relatively large and dome-shaped: the abdomen is very broad and pear-shaped, with the widest part towards the posterior end. The general colour is yellow.

Goniodes dissimilis resembles the preceding but is larger, from 2 to 2·5 mm. long. It is paler in colour. Both these species are found on the rump, and neck, and under the wings.

Bird-lice perish in a very short time if their host dies. They appear to have no power of migration such as the fleas, for example, possess. They are spread when two birds come into contact, either during 'treading' by the cock, or when poultry are closely confined. Species attacking the fowl are not found on ducks, and vice versa.

Mallophaga do not suck blood: they are incapable of so doing owing to the structure of their mouthparts. They live upon dead or nearly dead cuticular substances, upon the saddle hackles of feathers, etc. They set up irritation both on account of their biting habits, and also because of the scratching action of their claws as they move about. Badly infested fowls lose their feathers and become bare in patches. The constant irritation prevents the birds from resting, so that serious loss of condition results.

Control Measures.—Poultry kept under insanitary and overcrowded conditions are always worse sufferers from lice than birds kept in clean runs and houses with ample

30A - *Menopon pallidum.*

30B - *Lippeurus variabillis.*

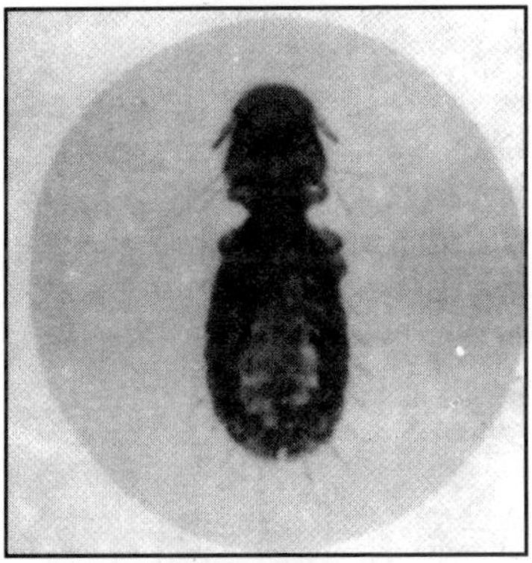

30C - *Goniocotes hologaster.*

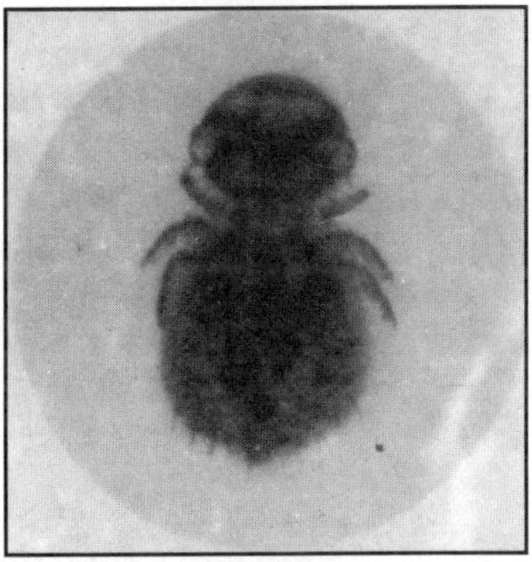

30D - *Goniodes dissimillis.*

FIG. 30. Bird Lice.

air, light and ventilation. To get rid of lice, dusting with pyrethrum powder or sodium fluoride has been found very efficient. All broody hens should be thus treated before being set. Particular attention should be paid to the areas round the vent, and under the wings. Young birds are sometimes greased against lice, lard, or a mixture of lard and sulphur, being applied to the head and neck, round the vent, and under the wings. Hens should have access to dry ground where they can make their own baths. These natural baths appear to be more efficient than the boxes of earth mixed with insect powder which are sometimes put into hen houses and runs.

Biting Lice which attack farm animals can be controlled by the same methods used against Sucking Lice (see p. 56).

SUB-ORDER.—2. SIPHUNCULATA.—Sucking Lice.

Many writers consider that the sucking lice constitute a distinct order, the Anopleura: as previously stated in this book, it has been decided to unite the Mallophaga and Siphunculata into one order.

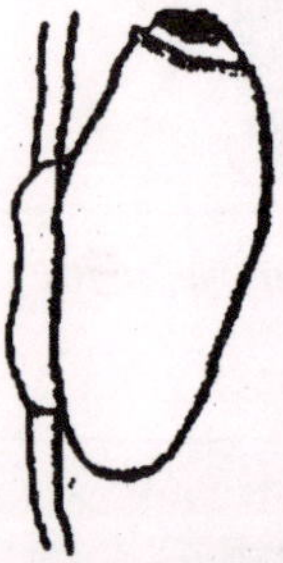

FIG. 31.—Egg or 'nit' attached to hair. British Museum Guide.

The sucking lice differ from the biting lice chiefly in the shape of the head and the structure of the mouthparts. The head is small in proportion to the rest of the body, and is longer than broad.

Mouthparts.—These are specially adapted for piercing and sucking. They are extremely small and difficult to investigate and when not in use they are withdrawn into the head.

Eyes are much reduced or entirely absent. Antennæ are three- to five-jointed. The thorax is small, with all its segments fused together. The tarsus bears a single powerful claw. The abdomen contains nine segments.

Wings are absent.

Metamorphosis is slight. In the case of the human Body Louse (*Pediculus humanus*) eggs, up to the number of 300, are cemented singly on to hairs by the female at the rate of eight or twelve a day. The egg, or nit, is about one-twenty-fifth of an inch long (Fig. 31). Under normal conditions eggs hatch in about a week, and the young louse escapes by forcing up the *operculum* or lid at the top of the egg. After three moults the louse becomes mature, this process occupying about seventeen days. Lice live from three to five weeks as adults.

There are at present about 120 known species of sucking lice, divided amongst four families. The two most important families are the *Pediculidæ* and *Hæmatopinidæ*. The Pediculidæ includes the loathsome human Body Louse already referred to.

Lice act as carriers of many diseases. They are the only carriers of typhus fever: they assist in the spread of relapsing fever, trench fever and plague.

Change of clothing and ordinary sanitary precautions absolutely prevent attacks by lice.

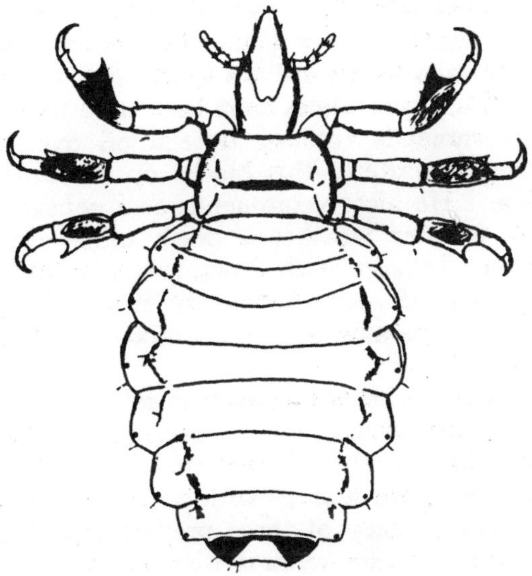

FIG. 32.—Pig Louse (*Hæmatopinus suis*).

The family Hæmatopinidæ includes species which frequently attack domestic animals. *Hæmatopinus eurysternus* infests cattle, chiefly along the middle of the back, on the loins, along the upper part of the neck and behind the head. *H. macrocephalus*, the female of which may be nearly 4 mm. long, plagues the horse. *H. suis* is very common on pigs, especially in the region round the ears. It is the largest of the common lice, females reaching 5 mm. (Fig. 32). *H. piliferus* infest dogs. Sheep and cats do not harbour members of the Siphunculata.

The presence of lice on farm animals is first indicated by the continual rubbing which is indulged in in order to allay the intense itching set up by the parasites. The irritation and loss of blood consequent upon an attack of lice leads to loss of condition, the coat becomes dull and ragged, and the skin becomes very dirty. In very bad cases the attacked animal may die.

Control Measures.—Clean animals do not become severely infested with lice. For this reason all stables, sties, loose boxes, etc., should be kept clean and disinfected by occasional lime washings.

To destroy lice which are actually upon the animals an application of some contact insecticide must be given, followed by a second application not sooner than five days and not later than eight days after the first. This ensures the death of any young lice which may have developed from nits which are not killed by these dressings.

Many substances have been used for killing lice. Leeney recommends rubbing linseed oil over the skin of cattle during winter: this blocks up the breathing pores of the lice. He also recommends as a contact insecticide for use in warmer weather a wash made up as follows:—Boil 4 ozs. tobacco with 8 ozs. soft soap in 2 gallons of water: syphon off and thoroughly shake the liquid with half a pint of paraffin oil. This mixture should be well brushed into the skin.

An infusion of quassia and soft soap makes a very efficient and safe insecticide to use against lice. It is made by boiling together approximately 1 lb. of soft soap and 1½ lbs. quassia chips with 2 gallons of water.

The efficacy of these preparations depends very largely upon the thoroughness with which they are applied to the skin of the affected animal. A stiff brush is the handiest and most efficient means of applying insecticides designed to kill lice.

Paraffin oil is sometimes recommended as a cure for lice. It is certainly very deadly to insect life in general, but it has a very irritating and harmful effect upon the skin of farm animals. Paraffin oil should never be used alone, but always diluted with other substances.

Lice on pigs can be removed by light applications of mercurial ointment round the ears and neck.

CHAPTER X

THYSANOPTERA—THRIPS

THE members of this order are well known as Thrips, Bladderfeet, Thunder Fly, Black Fly, or Flower Insects. They are minute animals, varying in length from one-third to one-fiftieth of an inch. In colour they are usually black, but yellowish and yellowish-brown species frequently occur. The body is slender, and sparsely covered with hairs. The antennæ consist of from six to nine joints. The chief characteristics of this order are the mouthparts, the wings and the feet.

Mouthparts.—The mouthparts differ from those of other orders of insects in being decidedly asymmetrical, or one-sided. They are adapted for piercing and sucking. The mouthparts are bent backwards underneath the head to form a short *mouth cone* (Fig. 33). The clypeus and labrum form the front part, and the labium the rear part of the cone. The sides are occupied by the galeal portions of the maxillæ, from which spring a pair of three-jointed maxillary palps. The labium projects beyond the upper lip and maxillæ, and the tips of both labium and maxillæ are provided with hooks. In the case of the Onion Thrips here figured, only the left mandible is developed, forming a long stylet: the right mandible is vestigial. The laciniæ of the maxillæ also form long piercing stylets, one on either side of a shorter central one, the modified hypopharynx. It appears that the insect when feeding first breaks through the cuticle of the leaf attacked by repeated jabs of the mandible, utilizing the weaker maxillæ to break down the softer mesophyll cells. The sap which exudes is sucked up through the muscular action of the pharynx.

Wings.—All individuals belonging to the Thysanoptera do not possess wings: some are secondarily apterous (see p. 14). Wings, when present, are very narrow and strap-shaped, and are remarkable in that very few veins are to be found: cross veins are almost entirely absent. Each

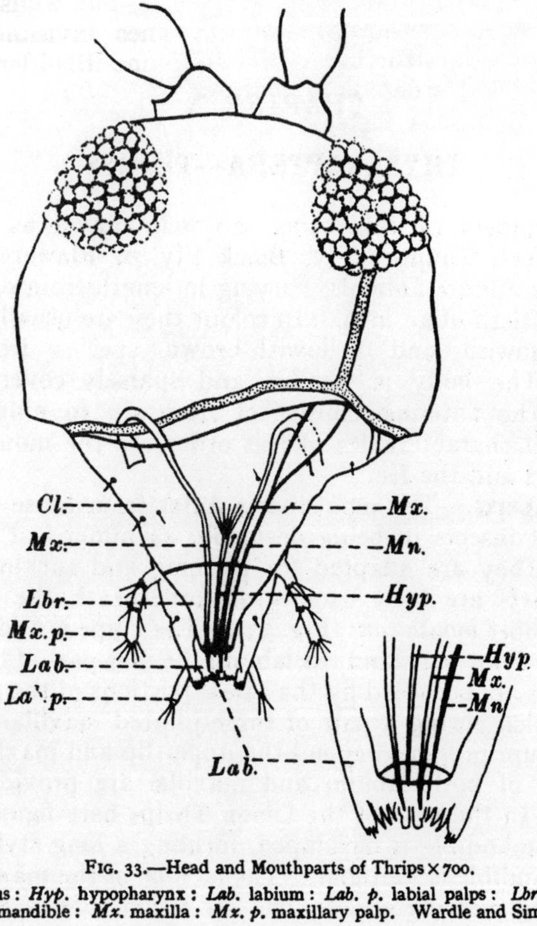

FIG. 33.—Head and Mouthparts of Thrips × 700.
Cl. clypeus: *Hyp.* hypopharynx: *Lab.* labium: *Lab. p.* labial palps: *Lbr.* labrum: *Mn.* mandible: *Mx.* maxilla: *Mx. p.* maxillary palp. Wardle and Simpson.

of the four wings is fringed with long hairs, and the front wing interlocks with the hind wing by means of a number of hooked spines near the base of the hinder one. When at rest the wings are laid parallel along the back.

Feet.—The tarsus of Thysanoptera has either one or two joints, the last segment carrying one or two claws and a curious bladder arrangement. This bladder can be protruded or retracted at will by the control of the blood pressure to the feet (Fig. 34). When protruded, the insect is able to cling to any kind of surface, but whilst resting the organ is withdrawn and becomes invisible. It is owing to this structure that the name Bladderfeet has arisen: the order is sometimes called Physopoda instead of Thysanoptera.

Metamorphosis. — The newly hatched thrip closely resembles the adult, both in appearance and habits. It moults four times. After the third moult the nymph becomes sluggish, and whilst not actually turning into a pupa, it behaves very much like one. Hibernation may take place in all stages. In some species males are very rare. Parthenogenesis frequently occurs. About one hundred species are known to occur in the British Isles.

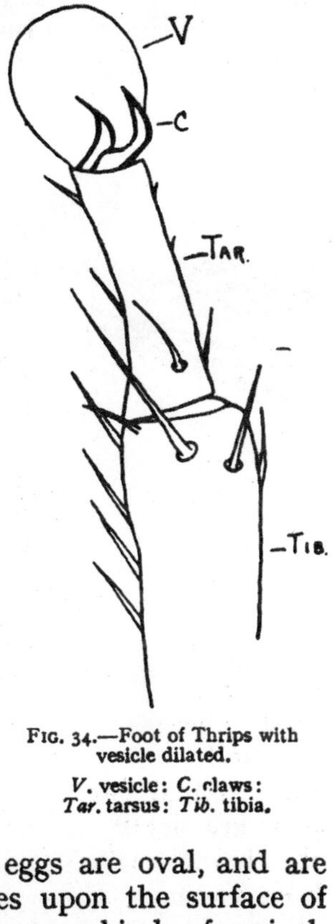

Fig. 34.—Foot of Thrips with vesicle dilated.

V. vesicle: *C.* claws: *Tar.* tarsus: *Tib.* tibia.

Thrips are widely distributed, being found in blossom of all sorts, on leaves, in decaying vegetation, and in the larger fungi. The eggs of one sub-order are somewhat kidney-shaped, and are laid singly in small slits cut in the plant tissue by the sawlike ovipositor of the female. In another sub-order the eggs are oval, and are deposited either singly or in masses upon the surface of leaves, etc. Thrips cause damage to many kinds of agricultural and horticultural crops and trees, and are frequently a serious pest in glass-houses. The best known species are those which attack pears, onions, peas, beans, and grasses.

When thrips attack flowers they frequently cause sterility, but it is also stated that they do considerable good in fertilizing beet and some other plants.

The Pea and Bean Thrips (*Frankliniella robusta*).

Description and Life History.—This insect frequently causes damage in the south and east of England to eating peas and broad beans. Adults are about one-twelfth of an inch long, black, with a lighter band across the thorax (Fig. 35). They appear from May to August: at first males are more common than females, but as the season progresses the males disappear.

Eggs are deposited in slits made in the stamen sheath, and they hatch in about nine days. The nymphs are orange in colour, but the last two abdominal segments are dark brown. After the first moult, the insect feeds for about three weeks before burrowing several inches into the soil. In this state the insect remains over winter.

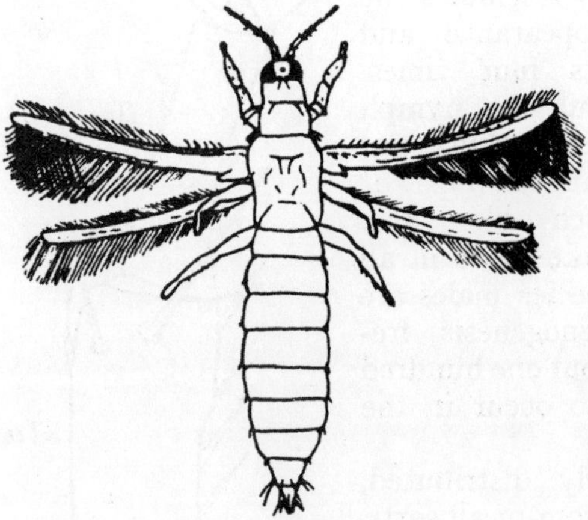

FIG. 35.—Female Pea Thrips (*Frankliniella robusta*). After Williams.

In spring two further moults occur and the adult emerges. There is only one brood each year.

Nature of Injury.—Damage is caused to young leaves in the terminal shoots, to flowers and to the pods. Attacked pods are sickly, under-developed, curled, and covered with the characteristic silvery brown areas where the larvæ have been feeding. These silvery patches are due to the presence of air under the epidermis, let in during the feeding of the larvæ.

Peas and beans grown on light land seem to suffer most severely, and no immune varieties are known at present. Early sown plants are less susceptible than late sown varieties. Attacks are most severe in dry weather, for rain kills enormous numbers of thrips.

Control Measures.—It was formerly recommended to burn old pea-sticks, but since hibernation takes place in the soil, this operation is quite useless. It must be confessed that beyond taking the precaution of not growing peas or beans on land close to that which suffered from thrip attack the previous year, very little can be done against this pest. Spraying is impracticable, and it is doubtful if soil fumigation would be an economical proposition except in the very worst cases. The lighting of smudge fires so as to allow smoke to drift over the plants has often had good results, provided the attack has been taken in time.

The Corn Thrips (*Thrips cerealium*) may cause damage to wheat and oats. Early in the year the insects live between the haulm and the sheathing leaf, and suck the plant juices, causing the stem to become twisted. Later they migrate to the ears, where they live and breed, considerably reducing the yield of grain. They hibernate as adults in the stubble, in soil, or on wild grasses. Control is difficult on account of the way the pest hibernates. Burning the stubble after a bad attack may be necessary, and the cleaning up of hedge bottoms reduces the amount of shelter.

Thrips in glass houses may be controlled by constant syringing with cold water, or by fumigation (see p. 270).

CHAPTER XI

RHYNCHOTA (HEMIPTERA)—BUGS

THE order of insects to which the Bugs belong is frequently called the Hemiptera, on account of the peculiar structure of the wings in certain families. It is preferable to use the term Rhynchota (Gr. *rhyngchos = snout*), because the most characteristic feature of the Bugs is the manner in which the mouthparts are drawn out into a long, sucking proboscis of a peculiar type. It is a feature which at once distinguishes the bugs from all other groups of insects, whereas the structure of the wings is very variable.

Mouthparts.—The snout is made up of elongated mouthparts, which at rest are carried along the ventral surface of head and thorax. There are upper and lower sheathing portions, probably corresponding to labrum and labium, whilst the more important organs take the form of fine stylets lying within the sheath (Fig. 36). This type of mouth structure is designed for piercing and sucking. Bugs, whether they be vegetable or animal feeders, obtain their food by piercing the tissues with their needle-like mouthparts. Into the wound they inject saliva, which by its irritant action ensures a copious supply of plant or animal juices, and may actually dissolve a certain amount of solid matter. The fluid is sucked up into the mouth partly by capillary attraction along the grooves of the stylet and partly by the action of the pharynx.

The Rhynchota exhibit great differences of size and shape. Some of them are very long, narrow insects (Capsidæ), some extremely broad and flat (Pentatomidæ), while others are soft-bodied and more or less pear-shaped (Aphididæ).

The Wings.—As a rule there are four wings, a pair on the mesothorax and a pair on the metathorax. One of the sub-orders (Homoptera) includes those families which have each pair of wings of a similar membranous texture, and the other sub-order (Heteroptera) those in which the mesothoracic pair differ from those on the metathorax.

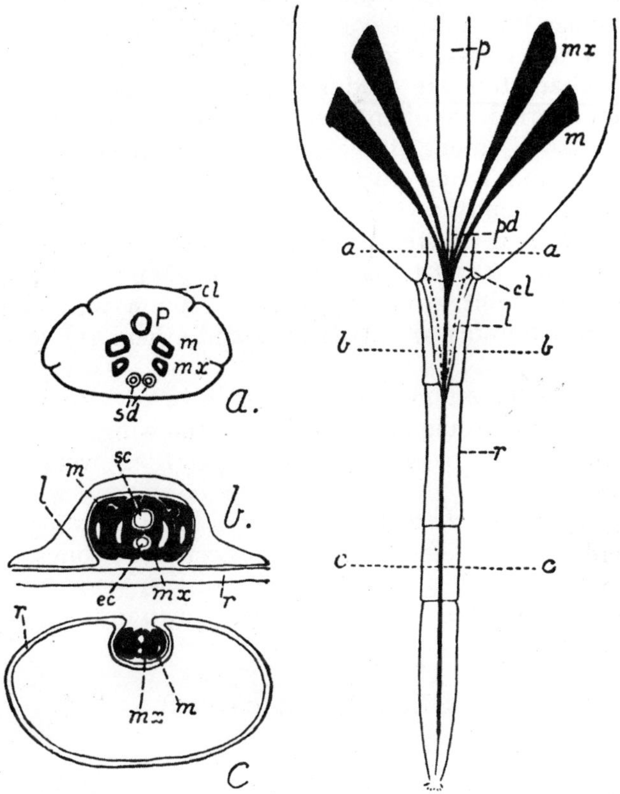

Fig. 36.—Mouthparts of a Bug.

On the left are transverse sections across the regions bearing corresponding lettering: the magnifications of these sections are not uniform. *cl.* clypeus: *ec.* ejection canal with salivary duct: *l.* labrum: *m.* mandible: *mx.* maxilla: *p.* pharynx: *pd.* pharyngeal duct: *r.* rostrum: *sd.* salivary ducts: *sc.* suction canal with pharyngeal duct.

From Imms, "General Text Book of Entomology."

The basal part of the mesothoracic wings is opaque and leathery in texture and only the tip is membranous. Such wings are known as *hemi-elytra*. The hind pair of wings is always membranous. Apterous forms are common among the aphides, and in certain species of scale insects

(Coccidæ) the males have only one pair of wings which is found on the mesothorax.

Metamorphosis is always slight. The egg on hatching gives rise to a nymph, which proceeds by a series of moults to become the fully developed adult. At each moult or ecdysis the wings become more and more developed and other characters are modified (Fig. 21). A resting, non-feeding period is often associated with the last moult. Particularly among the aphides, viviparous forms are common and living young are produced throughout the warmer parts of the year.

The order is of great economic importance both in this country and throughout the world. Many families are aquatic in habit, some are parasitic upon warm-blooded animals, a number are predaceous and others are vegetable feeders.

The Rhynchota may be conveniently divided into two sub-orders according to the nature of the wings—

1. Sub-order *Homoptera*—both pairs of wings membranous.

2. Sub-order *Heteroptera*—anterior pair of wings with chitinized bases (hemi-elytra): posterior wings membranous.

The chief families of the order are shown in the following table :—

RHYNCHOTA (ORDER)

Homoptera (SUB-ORDER)	*Heteroptera* (SUB-ORDER)
Family Aphididæ	Family Capsidæ
,, Coccidæ	,, Pentatomidæ
,, Psyllidæ	,, Cimicidæ
,, Aleyrodidæ	
,, Jassidæ	
,, Membracidæ	
,, Cercopidæ	

HOMOPTERA (SUB-ORDER)

FAM. APHIDIDÆ.

Aphides, Green Fly or Plant Lice.

Aphides are all small insects and their bodies are soft and not protected by heavy chitin plates. They are remarkable insects both in structure and habits. The family is world-wide in distribution and all species are vegetable feeders, constituting a severe menace to many cultivated plants. Most plants, both wild and cultivated, are attacked by one or more species of aphides, and during the summer months the insects live in colonies, often containing hundreds of individuals. One of the most remarkable characteristics of aphides is their method and rate of reproduction. In any summer colony, all the insects are females. Each of these females produces living young aphides, at a rate varying with temperature conditions and the food supply, but three or more are commonly born per day. Each one of these young aphides is a female, and in the course of a few days it becomes adult and produces more young aphides. This viviparous method of reproduction goes on throughout the warmer parts of the year and males are not usually found until the autumn. The summer method of reproduction is therefore parthenogenetic also and in a very short time huge numbers of aphides are produced. It has been said that the uninterrupted breeding of ten generations of aphides from a single viviparous parthenogenetic ancestor would produce a mass of organic matter equal to that of about five hundred million human beings (approximately the population of the Chinese Empire).

The aphides in a summer colony may be wingless (apterous) or winged (alate) (Fig. 37). The alate individuals migrate and spread the infestation to other plants. Males are usually not produced until the autumn and the fertilized female aphid then lays eggs. The winter is passed over as a rule in this stage. Such eggs may be laid on the plants or trees where the aphides feed in summer, or on a plant belonging to an entirely different botanical order. The winter eggs always give rise to viviparous parthenogenetic female aphides which are known as *stem-mothers*.

The sucking trunk or proboscis is usually well developed, and lies beneath the head and thorax in a horizontal position when at rest. When the insect is feeding the proboscis is forced more or less vertically into the plant tissue and sap is withdrawn from the sub-epidermal cells.

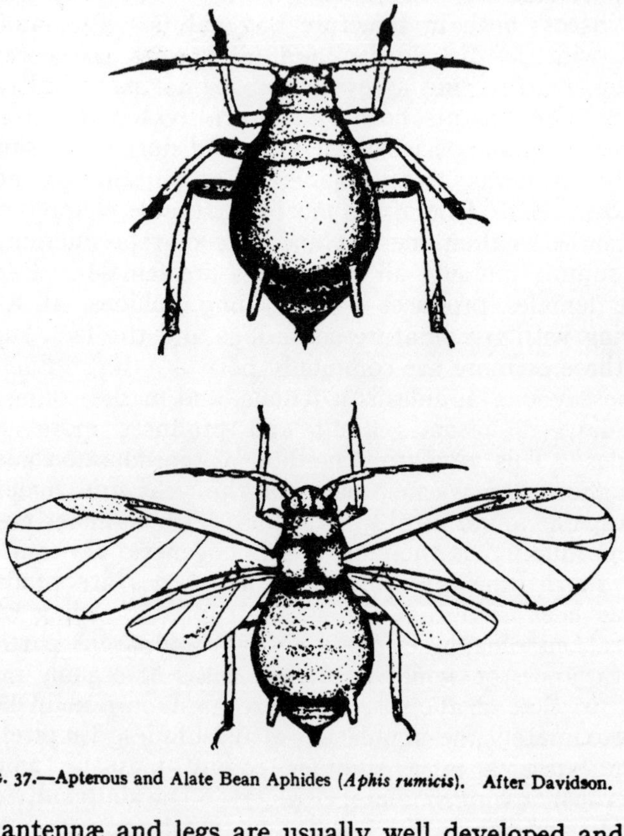

FIG. 37.—Apterous and Alate Bean Aphides (*Aphis rumicis*). After Davidson.

The antennæ and legs are usually well developed and the tarsus is commonly composed of only two joints. On the dorsal side of the fifth abdominal segment arises a pair of tubes known as *cornicles*. These vary in shape in different genera and their function is not known. Most aphides excrete a sugary liquid called 'honey dew,' upon which some species of ants readily feed. A waxy substance is also secreted from glands in the body, and this serves as a protective covering. It is most abundantly found on the Woolly Aphis and the Mealy Plum Aphis.

The damage done by aphides chiefly takes the form of a weakening of the plant due to loss of sap, and also to a poisonous secretion, an injection of which causes distortion. Large colonies of aphides may completely stunt young growth, and distortion of wood, leaves and fruit commonly follows on a bad attack. Moreover, black moulds grow in profusion upon the honey dew which is excreted, and the stomata or breathing pores of the leaves are thereby choked. There is evidence that the agent causing virus or degeneration diseases of potatoes, and other plants, is transmitted from diseased to healthy plants by aphides and similar bugs.

The family must be regarded as containing some of man's worst enemies, and the destruction might be greater were it not that aphides have a large number of insect enemies. Ladybirds (Coccinellidæ) and their larvæ feed readily upon aphides, as do the larvæ of the Lace-wing fly (*Chrysopa perla*) and the Hover flies (Syrphidæ). A great many parasitic Hymenoptera also utilize aphides as hosts, and they have been artificially introduced into some countries to control attacks.

APHIDES ATTACKING PLUMS AND DAMSONS

The Leaf-Curling Plum Aphis (*Anuraphis helichrysi = A. prunina = Aphis pruni*).

Description and Life History.—As is common among aphides, this species winters in the egg stage and the black shining eggs may be found at the base of buds on plums and damsons throughout the winter. As the buds are beginning to break in March or April, the eggs hatch, producing a purplish-coloured, wingless, viviparous female, the stem-mother. Young wingless viviparous females are quickly produced, greenish in colour at first, but becoming darker as they grow older, and they in their turn give rise to further generations. On hatching, the stem-mothers commence to feed on the young leaves which curl in a characteristic manner, and under this protection the colonies of young aphides are found. Later in the season the young wood growth is attacked, becoming severely distorted and twisted. During the early summer (June or July) winged forms are produced which leave the trees, so that after this period, though the trees may show evident signs of an attack, few if any aphides are to be found on it. It is not certain to what other plants these winged forms migrate, but similar aphides appear on Forget-me-not. During the autumn, winged forms again come back to the plums and from these arise the males and females, the latter laying the winter eggs.

Plants attacked and damage done.—All varieties of plums and damsons appear to be attacked, and the species is also found on blackthorn. Occasionally other trees, such as peach and apricot, are slightly damaged. The characteristic curling of the leaves is an obvious sign of the attack, and if this is severe, the leaves are so weakened that they shrivel, turn brown and die (Fig. 38). The young wood growth is also twisted and stunted and the fruit frequently drops off when partly developed. A severe attack may completely destroy the crop, and it is believed, moreover, that it also lessens the possibility of a good crop in the succeeding year, even if the aphis attack then is not severe. Like many other aphides, this species also gives rise to large quantities of sticky so-called 'honey dew,' which spreads over the surface of the leaves and quickly becomes covered with a black mould.

An attack is usually referred to as 'blight,' and one often hears that trees which are perfectly healthy suddenly become 'blighted' in a day or two, especially if there be a cold spell of weather at the time. The east winds of spring are invariably associated in the minds of many people with the appearance of 'blight,' and it is frequently averred that 'blight' is caused by some mysterious agent carried by such winds. Certainly the trouble often does appear very suddenly under these conditions, but the fact is that such climatic conditions merely check the normal growth and expansion of the leaves, while the aphides go on feeding and increasing in numbers, so that the damage rapidly increases and becomes more obvious.

Natural Enemies.—A number of insects prey upon this aphis, especially ladybirds and their larvæ (Coccinellidæ), a number of parasitic 'flies' of the order Hymenoptera,

FIG. 38. Plum shoot attacked by *Anuraphis helichrysis*

the larvæ of Syrphid or 'hover' flies and the lace-wing 'fly' larva. These check the increase in numbers to a very great extent, but do not usually appear until the aphides are very plentiful in the early summer, by which time the damage h been done.

Control Measures.—The only effective means of control

is by spraying, and two types of spray fluids may be employed.

1. A 'contact' poison (see p. 250) in which Nicotine is the most common toxic substance used, combined with soft soap as a 'spreading agent.' Certain preparations involving the use of vegetable poisons such as Derris root are also effective. These washes must be applied as soon as the first signs of attack are seen and before the leaves become heavily curled.

2. Egg-killing Winter Washes, generally known as Tar Oil Washes. These are applied during the winter when the buds are perfectly dormant, and they possess the power of destroying the eggs on the tree. If they are properly used the necessity for summer spraying is obviated.

The Mealy Plum Aphis (*Hyalopterus arundinis = H. pruni*).

Description and Life History.—The eggs of this aphid are similar in appearance to those of the preceding one and are laid at the bases of the buds. They hatch much later in the year than those of *A. helichrysi*, normally not until late April or early May, and they produce stem-mothers, from which succeeding female generations arise. Winged forms are produced late in June, and of these some, at any rate, migrate to reeds and grasses, but some remain on the plum. A return migration again takes place in the autumn. This insect is light green in colour, but produces whitish flakes of a waxy or mealy substance by which it is easily distinguished. The whole under side of the leaves is often covered by these aphides, but little or no leaf curling takes place.

Plants attacked and damage done.—All kinds of plums are attacked, and the aphis is also found at times on apricots, peaches and nectarines, as well as on reeds and grasses. Attacked leaves become very brittle, and though there is very little leaf curling and distortion of young wood, the tree suffers from loss of sap. A very unsightly appearance is given to the tree, since large quantities of honey dew are produced.

Natural Enemies.—Similar to those of *A. helichrysi*.

Control Measures.—1. For summer spraying a paraffin

emulsion is sometimes used, but summer spraying may be disappointing because the wash is unable to penetrate the waxy covering of the aphis.

2. The tar oil egg-killing washes would probably be effective, though at present little is known of their effect on this particular species.

The Hop Damson Aphis (*Phorodon humuli*).

Description and Life History.—This aphis, which is green in colour, is found on both damsons (and sometimes plums) and hops. It appears that there are two definite life-cycles which may be pursued, one entirely on hops and one in which a migration between the hop and the damson takes place.

In the life-cycle which is confined entirely to the hop, the aphis hibernates as a wingless viviparous female in the soil close to the plant, which she ascends in spring and on which young aphides are produced. Towards the end of the year some winged individuals of both sexes appear, the females of which are responsible for migration to damsons, the wingless ones remaining on the hops. The second kind of life-cycle, therefore, may be looked upon as starting with these fertilized females, which lay the black winter eggs on damsons. These eggs hatch into stem-mothers in spring and a viviparous and parthenogenetic method of reproduction goes on until winged forms appear, which migrate to the hop again.

Plants attacked and damage done.—The damson (and various plums) and the hop are the host plant of this aphis. It is chiefly responsible for damage to the hop, where leaves and the tips of young growth are attacked and weakened by the extraction of sap, and covered with honey dew. The black moulds which grow upon this lead to the term 'black blight' being applied to the attack on hops. Invasion of the hop cones also takes place later in the season, with the result that these may be severely depreciated in market value. The earlier stages of attack lead also to a great diminution in the crop.

Natural Enemies.—The insects previously mentioned as preying upon aphides also attack this species, but seldom do more than to help keep it in check.

Control Measures.—On the hop a nicotine and soft soap wash is frequently employed for summer spraying, while on the damson a similar wash is effective. Tar oil washes also control the insect on damson by destroying the eggs laid there.

APHIDES ATTACKING THE APPLE

There are a number of species found on the apple, of which four are of some importance. Since the life histories of three of these aphides are all similar and present no exceptional features, they will be considered together. The fourth species, known as the Woolly Aphis, must be dealt with separately.

The Permanent (or Green) Apple Aphis (*Aphis pomi*).

Description and Life History.—The apterous females found on the apple are yellowish green in colour and the whole life history is gone through on the tree. The eggs are black and laid in large clusters, particularly on young growths and on young trees. These eggs hatch in April, giving stem-mothers from which succeeding generations arise. Males and females (the males are wingless in this species) are produced late in the season and the winter eggs deposited.

Plants attacked and damage done.—Except to the tips of young growth, where this aphis normally feeds, little damage is done as a rule. There is little or no leaf curling or distortion such as is caused by the Rosy Aphis. There is usually no migration, though winged forms are produced in summer, but the species has been found on thistles.

The Grain and Apple Aphis (*Rhopalosiphum prunifoliæ*).

Description and Life History.—This is a somewhat delicate green aphis with darker green markings on the back. The eggs, deposited singly and not in batches, are found in winter on apples and from these the stem-mothers hatch. Winged migrant forms are produced very early and these migrate to cereals and grasses, from which return migrants come in the autumn, the females depositing winter eggs.

Plants attacked and damage done.—The presence of this aphid on cereals is not usually accompanied by any appreciable damage. On the apple, the early generations invade blossom trusses and feed there, causing at times appreciable damage to the delicate flower-stalks. Wood and leaf growth is not attacked and no curling or distortion is caused.

The Rosy Aphis or Blue Bug (*Anuraphis roseus*).

Description and Life History.—The name ' Blue Bug ' is commonly applied to this aphis because of its prevalent bluish or purplish appearance. Some individuals are reddish in colour, but all are of a swollen or globular appearance. In winter the eggs may be found on apple trees, and they are laid more or less singly and not in clusters. The stem-mothers arising from these eggs produce young, and colonies may rapidly develop into large proportions, covering the under side of the leaves and spreading to the young wood growth. Later they attack the young fruit. During June and July winged forms are produced which leave the apple, but their destination is uncertain. They are found in America to go to plantains, but so far this has not been found to be the case in England. Some have been found on the guelder rose, and a very similar, if not the same, aphid on the roots of docks. At all events, during the autumn a return of winged males and females takes place and the eggs are laid again.

Plants attacked and damage done.—Only the apple suffers damage from this aphis, and the attack may be very severe. First the young leaves are attacked and these curl up. Pinkish or yellow blistered patches are formed under which the aphides shelter. Often the leaves are completely shrivelled and killed, and covered with honey dew. Young wood which is attacked becomes twisted and distorted, and the fruit remains wrinkled and dwarfed, a good deal of it falling before it develops. The effect of a bad attack may be seen in the following season and often for a longer period, since it takes the tree a long time to recover from so severe a check. A similar aphis, *Anuraphis cratægei*, also produces pink or red coloured patches on the foliage of apples.

Natural Enemies.—All these three apple aphids are subject to attacks by parasites of the order Hymenoptera, and by ladybirds, lace-wing flies, and syrphid flies. They are not more than checked, however, by these insects, and in spite of them trees which are not treated become increasingly badly affected.

Control Measures.—As is the case with the plum aphides, 'contact' insecticides usually involving the use of nicotine or derris root extract are effective for summer spraying. For Rosy Aphis, however, it is absolutely essential that spraying shall be done before the leaves curl, for once curling has taken place the spray fluid cannot be made to reach the aphis in these protected situations and little good will be done. An effective control can be obtained by using a coarse driving spray soon after the insects have hatched.

Tar oil winter washes give an excellent control, and trees so treated need not be sprayed in the summer if the winter operation is done well with a good wash.

The Woolly Aphis (*Eriosoma lanigerum*).

This insect is common wherever apples are grown, being now world-wide in distribution. It is frequently known as American blight, though its original connection with that continent is obscure, it having been known in this country since 1776.

Description and Life History.—The true colour of this insect is not usually apparent, owing to the large masses of white woolly material with which the individuals are covered. It is this woolly appearance which is so obvious a sign of the presence of the insect during summer, when large patches of it are to be seen on neglected trees. The aphis itself is purplish in colour and secretes the long threads of waxy material from glands in its body.

The life history is somewhat complicated by the fact that some of the insects remain on the apple over winter as wingless viviparous females, but there is a migration to the elm tree, on which the winter egg is laid. The female on the elm lays a single egg only. Further, the root system of the apple harbours colonies of this aphis as well as the aerial parts of the tree (Fig. 39), and a migration of wingless females takes place in the autumn from the branches to the roots, and in the spring back again from the roots to the branches. Throughout the year, however, some aphides may be found in both situations; the migration is not a complete one. Throughout the summer, from mid-June onwards, the typical woolly colonies are found on the branches of apple trees. These consist entirely of wingless viviparous females, but later in the year a number of winged viviparous females are produced which fly away to the elm. There they give rise to wingless generations of both males and females, the latter each depositing one egg in a crevice of the bark. This egg hatches in the

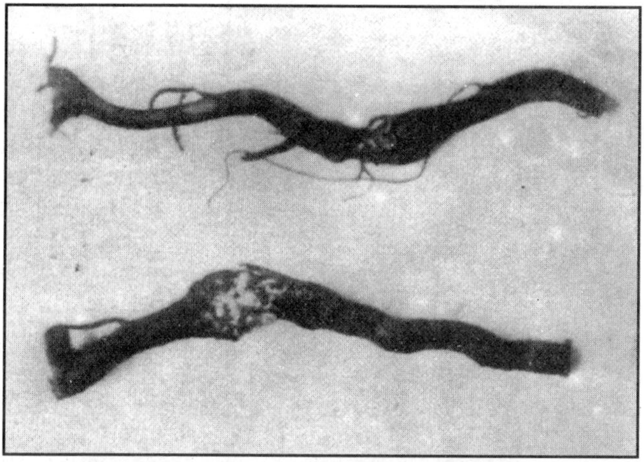

FIG. 39. - Root galls produced by wooly Aphis.

following spring, giving rise to a stem-mother which produces a small colony of wingless viviparous females. These feed on the leaves of the elm, principally on suckers, and in late June or early July, a winged female generation is produced, the individuals of which leave the elm and some, at all events, go to the apple, where they continue the viviparous form of reproduction. Except during the late summer, therefore, Woolly Aphis is distributed on two trees, apple and elm, and only during late summer and early autumn are all the insects confined to the apple.

Plants attacked and damage done.—The only plant of economic importance to which direct injury is caused is the apple, and apart from the generations found on elm, the 'woolly' form has been found on hawthorn and a similar if not identical insect on pear. The injury is chiefly caused to the wood in apples, though in very heavy infestations, colonies may form on the leaves and even fruit. The effect of the aphis on the bark is very marked. Large swollen galled areas appear, accompanied often by cracking of the bark, and in such hiding-places the aphids congregate (Fig. 40). They develop year after year in old scars of this nature, and the wounds so caused afford a place of entry for the Apple Canker fungus, which often 'rings' the branch completely; killing out all wood above the point of attack. The injury to trees is mainly due, however, to the loss of sap caused by the sucking of the aphides, young trees being almost crippled if the attack is allowed to persist.

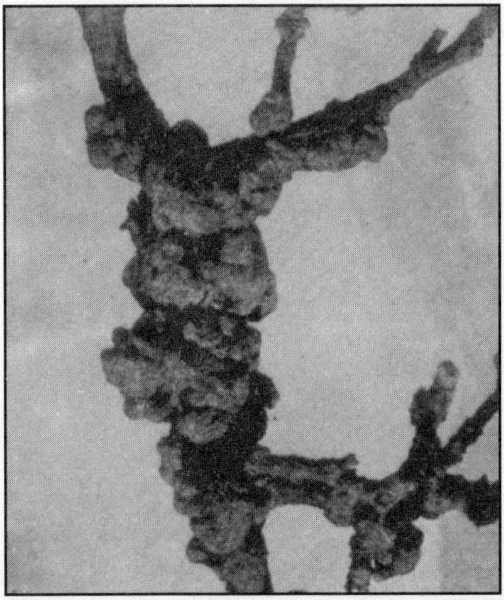

Fig. 40. - Branch of apple with swellings caused by Wooly Aphis.

Freshly cut surfaces form suitable points for a Woolly Aphis attack, and in consequence pruned or broken boughs, and the points where grafts are inserted, are frequently found to be attacked.

Natural Enemies.—Apparently owing to their woolly wax secretion when on the apple, they are relatively free from attacks by such insects as ladybirds and hover-flies which are so largely responsible for keeping other aphides in check. A few of the parasitic Hymenoptera prey upon them during the summer, and in winter various birds, especially tits, seek out concealed individuals and do much good in feeding upon them. On the elm, hover-flies and some Hymenoptera are found attacking the generations which are produced early in the year, and though some reduction in the migrant forms is undoubtedly brought about thus, the insect has few really effective enemies.

Control Measures.—Owing to the fact that at any given time both aerial and root colonies of woolly aphis exist on the apple, it is obvious that measures taken must include those in both situations. In the case of young trees, such as those sent out from a nursery, effective control can be brought about by fumigation with hydrocyanic acid gas (see Chap. XIX). It is most important that trees should be clean before they are planted.

Root Colonies.—On young trees the insects living on the roots can be destroyed by fumigation and the trees may be lifted and treated in this way. In the case of older trees injections of carbon disulphide are usually made in the soil around the tree in the following manner. Four holes are made with a crowbar around the tree and 2 feet away from the trunk. These should be about 6 inches in depth, and into each hole one fluid ounce of carbon disulphide is poured, the hole being immediately covered firmly. Carbon disulphide gives off a heavy poisonous vapour which penetrates through the soil and kills off the insects. The liquid must not be allowed to come in contact with roots, but in gaseous form it does them no damage. There is no necessity to treat the soil farther from the tree than 2 feet, since on old trees the colonies of aphides are found only on large roots fairly close to the trunk. The quantity of carbon disulphide given above is suitable for a well-grown tree fifteen to twenty years old. A special soil injecting pump is made for use with carbon disulphide where this work is carried out on a large scale.

Aerial Colonies.—On young trees, where there are comparatively few in a garden, they may be treated individually by hand, using a stiff paint brush and methylated spirit. The spirit must be well worked into scars and cracks so as to touch every aphis, and if this is done as soon as colonies appear the tree can be kept free. Young trees should on no account be allowed to become badly attacked before steps are taken to clean them. On older trees, individual colonies on a few trees may also be dealt with by using a paint brush and paraffin, but it is not wise to use raw paraffin on young trees or injury to the bark may result. A winter wash of caustic soda or one of the brands of tar oil washes will clean out stragglers which are over-wintering, and the latter washes also seem to prevent such severe summer infestations. This may be due to the fluid which runs down into the soil, where possibly it is fatal to the root colonies and so prevents much upward migration in spring. Grease banding has been advised as a means of stopping this upward migration, but it is exceedingly difficult to grease-band old trees effectively. There are usually crevices by means of which aphides can pass underneath the paper. On a large scale a late washing with a contact insecticide is probably the best means of control. Such a washing should be done immediately the apples have been gathered, a time when there will have been little or no migration either to the roots of apples or back to the elm. A number of other aphides will also be killed in the process before they have laid their winter eggs. For this purpose a fairly strong paraffin emulsion may be used, as there will be no danger of damaging the foliage at this time of year.

Resistance to Attack.—It has been found that not only do varieties differ in their resistance to the attack of Woolly Aphis but that certain apple stocks possess an almost complete immunity to root attacks, so that this form of damage does not occur when susceptible varieties are grafted on to such stocks. Northern Spy and Winter Majetin stocks are the two best known examples of this, and, although not widely used in this country, they are employed abroad, particularly in Australia, where they have found great favour. Of the varieties which are susceptible to attack, soft-barked kinds such as Cox's Orange Pippin, Warner's King, Blenheim Orange and Worcester Pearmain, among others, are most frequently infested, whereas Newton Wonder and Bramley seedling seem to be less attacked.

Note.—By the Sale of Diseased Plants Order, 1921, any

person is prohibited from selling, for planting, trees which are appreciably infested with this insect, under penalty of a fine of £10.

CURRANT AND GOOSEBERRY APHIDES

A number of distinct species of aphides are found on currant and gooseberry bushes, of which five are of some importance. The complete life-cycle of some of these has not been traced out. All except one feed on the aerial parts of the bushes, the exception being a root aphis on currants. Very often all the species described below may be found on the same bush, and little can be learned of the species which are present by examining the type of damage. Curled masses of leaves covered with honey dew simply show the presence of aphides and gives no clue as to species. The most common leaf and shoot feeding species are as follows:

Aphis grossulariæ.

This is a green plump aphis, which may, however, vary in colour, individuals of a slaty grey colour being sometimes found. Though chiefly attacking gooseberries, it is found on currants as well. The winter is passed on the bushes in the egg stage. Stem-mothers hatch from the eggs in April and the succeeding generations live on the young shoots and underneath the leaves. Winged individuals which are produced in the summer migrate to lettuce, sowthistle and other plants, and a return takes place in the autumn when the winter eggs are laid.

The injury caused to the young shoots and leaves may be very severe and the bushes at times are almost defoliated. Young growth is twisted and stunted and the leaves, which show pale or reddish blistered patches, are grouped together in a mass and covered with honey dew. Apart from weakening the bush, severe attacks may result in a great deal of the fruit falling.

Myzus lactucæ.

This is another green aphid which has dark lateral marks on the body. The eggs are laid on currants and gooseberries and over-winter there, stem-mothers hatching in the spring. The early generations are found on the under side of the leaves, and the winged migrants produced in summer fly to the lettuce and other plants. A return migration takes place in autumn, the females depositing winter eggs.

The leaves attacked by this aphis curl up into tufts, but these are generally looser in construction than is the case with the previous species.

The Hairy Currant Aphid (*Capitophorus ribis*).

Pale yellowish green in colour and fragile in form, this species may be distinguished from other aphides on currants and gooseberries by the numerous hairs with which it is covered. These hairs have swollen ends (*capitate*), a feature which gives the aphis its name. Its life history is similar to that of the preceding species, and migrant forms fly to dead nettles and *Stachys*.

These aphides are commonly found under the red blisters which are so often seen on red and white currants. Whether or not they are responsible for the blisters is not certain, as they may also be found feeding on the leaves elsewhere. The leaves are not markedly curled or tufted.

The Currant and Lettuce Aphid (*Amphorophora cosmopolitana*).

This species is also green in colour and is very common. Its life history is exactly like those already described on currants, and the migrant forms fly in summer to lettuce and sowthistle.

Loose, curled tufts of leaves are characteristic of the attack of this aphid.

Natural Enemies.—Ladybirds and their larvæ, the larvæ of lace-wings and syrphid flies and hymenopterous parasites all exact their toll of these aphides.

Methods of Control.—1. Tar oil washes give some measure of control, but the eggs of many of these aphides are laid in sheltered places in crevices of the bark, so that unless spraying is very thoroughly done the control obtained by this method is apt to be disappointing.

2. Spraying with a contact poison such as nicotine is very effective if done before leaf-curling takes place.

The Currant Root Aphis (*Schizoneura ulmi*)

Like the Woolly Aphis on apple, this aphis which is found

on the roots of currants and gooseberries during the late summer and autumn, is covered with a woolly material. Winged generations leave the elm during late summer and establish themselves on currant and gooseberry roots, a return migration taking place again sometimes in the same autumn, but more commonly during the following spring. Full details of the life-cycle are, however, unknown.

On the elm this aphis causes considerable curling of the leaves, while the gooseberry and currant may suffer severely from the root-attacks. Little is known, however, about the prevalence of this pest. It is almost certain that it is more common than has been supposed, and if this is the case then it is probable that it does considerable damage.

No adequate information is as yet available on the possibility of root fumigation or other measures.

OTHER APHIDES OF IMPORTANCE

The Black Bean Aphis, Black Fly, Collier, Black Dolphin, etc. (*Aphis rumicis*).

Description and Life History.—As the name indicates, the prevailing colour in this aphis is black, and its wide distribution and familiarity to most agriculturalists and horticulturalists has earned it many local names. The winter eggs of this aphis are found on the spindle tree (*Euonymus europæus*) usually around the bases of buds and in crevices of the bark. These eggs hatch in April when the buds on the spindle tree are developing, and give rise to wingless stem-mothers. The progeny of the stem-mothers are all females. In a generation or two some winged individuals are produced which migrate to the summer host plants (Fig. 37). Among such plants are beans, thistles, docks, poppies, furze, fat hen (*Chenopodium*) and shepherd's purse (*Capsella bursa-pastoris*). On these plants the winged females give rise parthenogenetically to another wingless generation, and these in turn reproduce in the manner usual among summer generations of aphides. Small colonies, however, may be found on *Euonymus* throughout the summer. On such plants as beans, reproduction takes place at a great rate. Each aphis, under favourable climatic conditions, produces about five young daily, and this may go on for many days until the progeny of one aphis amounts to forty or fifty. Each of these young female aphides reaches maturity in a week or so, and in its turn reproduces. This means that if all the individuals from one single parent survived and reproduced, there would be at the end of five generations over 100 million aphides. It is obvious, therefore, even allowing for the numbers which do not reach maturity, that the colonies of these insects grow at a very rapid rate. When this results in overcrowding, winged forms, all females, are produced which fly away to other plants and spread the infestation. Eventually, in September, winged males and females are produced, the females being incapable of viviparous reproduction. These forms migrate back to the spindle tree, where the fertilized females deposit their winter eggs.

Plants attacked and damage done.—Both broad and runner beans are frequently attacked, the infestation taking place at the tips of the shoots where large colonies are formed. The weakening of the plant is very apparent, and owing to the loss of sap the flowers may fail to set and many of the young pods fall off. Mangolds and sugar beet have also been found seriously infested with this aphis, the leaves becoming curled and distorted, and in bad cases the plant seriously checked. A great deal depends, however, on weather conditions, which considerably influence the rate of reproduction of the aphis.

Natural Enemies.—The insects previously mentioned as attacking aphides on other plants do much to keep the numbers in check.

Control Methods.—On a farm little can be done when the aphis attack occurs under conditions favourable to its increase. Well-established bean crops, especially when winter sown, appear to be less infested than others, and varieties of the Longpod and Mazagan type are also found to be more susceptible than the Tic beans.

In the garden the task is easier, for if the growing tips of the plants are pinched out as soon as any infestation is seen, the attack does not as a rule become severe. Spraying with a contact insecticide is also quite effective.

It might be supposed that eradication of the spindle tree would remove the possibility of attack altogether. While this might be useful to some extent, it would appear that the problem is not so simple, since in the south of England, where *Euonymus japonicus* is grown as an evergreen garden shrub, colonies have been found in winter on this plant. The ordinary type of parthenogenetic reproduction may also be carried out throughout the winter, by small colonies of aphides living on the young leaves of mangels growing on rubbish heaps. There is evidence also that manuring has an effect on the infestation, or at any rate on the rate of reproduction, but so little at present is known of this that it is unwise to do more than mention the fact. In pot experiments it has been found that a complete manuring, especially the addition of soluble potash and phosphates to the soil, has the effect of increasing the rate of reproduction, though on a field or garden scale one must not forget to interpret the results of treatments of this kind in view of the ultimate crop obtained. The benefit which the plant derives may conceivably outweigh the adverse effect due to the increase of the aphis.

A large number of other species of more or less importance occur on various plants such as the brassicas, cereals, and potatoes. Particularly in the case of potatoes a good deal of attention has recently been given to aphis attacks because of the probability that some of the so-called 'degeneration' or 'mosaic' diseases of potato may be spread from infected to healthy plants by means of sucking insects. The 'virus' causing these diseases seems to be capable of being transmitted from a diseased to a healthy plant through the sap sucked up by the proboscis of the aphis, re-infection being carried out when the insect injects saliva into a fresh puncture.

FAM. COCCIDÆ.

Scale Insects and Mealy Bugs.

These very remarkable insects are reckoned among the worst pests in the world, a great number of species being responsible for damage to plants. In this country, however, there are relatively few which are of importance. The scale insects are so called because of the fact that there is formed over the back, and completely covering the insect, a hard, waxy scale which affords it protection. These scales vary in shape and size according to species, and they, rather than the insects themselves, are most often seen. The mealy bugs derive their name from the white mealy substance with which they are covered, very much like that associated with the Mealy Plum Aphis.

The Coccidæ have a piercing and sucking proboscis, by means of which they extract the sap of the plants on which they feed. There is a marked difference between the males and females both in appearance and life history. The females are by far the more common, and are wingless and very degenerate in many ways, living for the majority of their lives under the scale which they have formed. They look very little like insects at all, for segmentation is often not at all apparent, while the antennæ and legs are frequently very small or almost wanting.

Males in many species are seldom or never produced, a parthenogenetic method of reproduction going on apparently indefinitely. Mouthparts also are wanting in the males, which do not feed and are very short-lived. They are frequently winged, but possess only two wings (mesothoracic), and undergo what is practically a complex metamorphosis, a true pupa being formed before the fully developed male is produced. This metamorphosis is, of course, very unusual for the Rhynchota.

Some of the Coccidæ are of importance because they provide substances which have been turned to use by man. The red dye known as cochineal is prepared from the dried females of a scale insect which lives on cactus plants in Mexico. This insect is *Dactylopius coccus*, frequently called the cochineal insect. The resinous substance from which shellac is made is a product of an Indian species, the 'lac' insect, *Tachardia lacca*.

The Mussel Scale (*Mytilaspis pomorum* = *Lepidosaphes ulmi*).

Description and Life History.—On the stem and larger branches of neglected apple trees may frequently be seen large numbers of small brown objects about one-eighth of an inch in length, shaped very much like a mussel shell (Fig. 41). These are the scales which cover the insects themselves, and in winter and early spring, if one of these scales be carefully lifted up with a needle, a number of tiny white eggs may be seen underneath. During late May or early June these hatch, and tiny flat, wingless nymphs emerge and work their way out from under the scale. For a few days they move about on the tree, but soon settle down, drive in their proboscis and remain fixed in that position for the rest of their lives, if females. In this early mobile stage they may easily be carried to other trees by wind or by clinging to the hairs of other insects. As soon as they have attached themselves firmly to the bark, they begin to produce the scale, under which they develop until they are fully grown, but even then

FIG. 41. - Apple tree with Mussel Scale (*Mytilaspis pomorum*).

they possess few of the characters by which an adult insect is recognized. The very great majority are females, and in the late summer they lay their eggs and die. Males are very rare indeed on fruit trees, and year after year the insect may continue without producing any male forms at all. When males are produced they form smaller, straighter scales than females, and they pass through a pupal stage before emerging as adults. The adult male has wings only on the mesothorax. Its mouthparts are not developed at all, and it is very short lived.

Plants attacked and damage done.—This scale is found on a large number of different trees, but chiefly on apples, though pears and currants may be attacked. It is more abundant on old neglected trees than on young stock, and though it does not appear to cause very serious injury to old-established trees, these provide a source of infection. Young trees are very severely crippled by heavy attacks. Trees should be carefully examined for these insects before they are planted and a watch kept at all times so that any tendency to increase can be checked. The damage is caused by the loss of sap which these insects extract.

Natural Enemies.—Such birds as tits and tree creepers search for and devour the scales and eggs, especially during severe weather. Some parasitic Hymenoptera attack them but only a slight control is effected in this way.

Control Measures.—Unless trees are seriously neglected, they seldom become heavily infested, and routine spraying will generally keep down the numbers of scale insects to a negligible quantity. Winter spraying is the most satisfactory form of control, strong paraffin emulsions or tar oil washes being effective if the tree is thoroughly wetted. The scales are softened by a caustic soda wash also, and may easily be removed after such an application by means of a stiff brush.

Towards the end of May or early June a contact insecticide will kill the young unprotected nymphs just after they hatch and before they are effectively covered by a scale. It is most important, however, to keep down infestation rather than to allow it to become severe before measures are taken against it.

The Oyster Shell Scale (*Aspidiotus ostreæformis*).

Description and Life History.—This scale is much the size of a pin-head, greyish or greenish in colour and more or less circular in shape, deriving its name from the similarity it bears to a tiny oyster shell. The eggs, which are found under the scales in summer, hatch out, and the nymphs, after a short wandering period, settle down, pierce the bark and form a small covering scale. The winter is passed in this stage and not in the egg stage as with the mussel scale. Males are produced in the early summer, and these pair with the females, after which the eggs are laid. There seems to be only one generation annually in this country.

Plants attacked and damage done.—A large range of plants, including apple, plum, pear, cherry, apricot, nectarine and peach are attacked and occasionally currants, while the insect is also found on heather.

The loss of sap occasioned by the feeding of the insect has a weakening effect on the tree in the same manner as with the mussel scale.

Natural Enemies.—The larvæ of a ladybird feed to a large extent on this scale: possibly birds also play some part in reducing its numbers.

Control Measures.—It is seldom that the Oyster scale becomes serious where routine spraying is carried out, and winter washing helps to keep trees free from it.

Lime sulphur, used at winter strength, is effective against it. Trees should on no account be purchased which have this scale on them. The pest is scheduled under the Sale of Diseased Plants Order, 1921 (see Woolly Aphis).

The Brown Scale (*Lecanium persicæ var. ribis*).

Description and Life History.—The scale which covers this insect is a large one, more or less circular, but usually rather longer than broad. It is about one-eighth of an inch in diameter, very convex and of a deep reddish brown colour when old. At first it is of a lighter yellowish tint with darker bands of colour on it.

Eggs are laid during the late summer, hundreds being produced by a single female. The body of the insect is very soft and swollen, and when she dies this hardens into the protective scale under which the eggs are found (Fig. 42).

The young nymphs emerge through an aperture at the posterior end and obtain cover by hiding under rough bark, where they pass the winter, appearing as tiny reddish-brown specks. With the spring they become active for a time and develop by early summer into adults. The males of this species are unknown. One generation only per year occurs in this country.

Plants attacked and damage done.—Gooseberry and currant bushes, and peaches and nectarines are most commonly attacked, though the insect is found on many other plants. Like most of the scale insects it is most abundant in sheltered, warm situations, and wall trees are particularly liable to attack, as well as peaches, etc., under glass.

FIG. 42 - Brown scale of Currant (*Lecanium persicae*)

Early ripening of the foliage and very small crops of fruit are associated with bad attacks of this scale, due to the extraction of sap. The older branches are more frequently attacked than young wood.

Control Measures.—Few, if any, natural enemies of this scale materially assist in checking it. Winter washing is effective if thoroughly carried out, so that the fluid reaches the young nymphs in hiding, and to facilitate this rough bark should be removed as far as possible. Lime sulphur, caustic and paraffin washes may be used, and possibly tar-oil washes would be effective, though little definite information is available concerning their use on this pest.

FAM. PSYLLIDÆ.

Jumping Plant Lice.

Most of the members of this family are small insects whose power of flight is not well developed, although the wings are fully formed. They are characterized by short jumping flights, the stout hind pair of legs being adapted for this purpose, but they appear to be incapable of remaining in the air for any long period.

The immature (nymph) stages usually exude quantities of a viscous translucent oily substance in globules, and white waxy threads of a similar material are also produced. These nymphs are very flattened in form and are responsible for the injury to plants. Little or no direct damage is caused by the adults.

In this country species occur on apple, pear, and box: the one on apple may be taken as typical.

The Apple Sucker (*Psylla mali*).

Description and Life History.—The eggs of the apple sucker may be found on trees throughout winter and early spring, firmly attached to the bark by means of a small

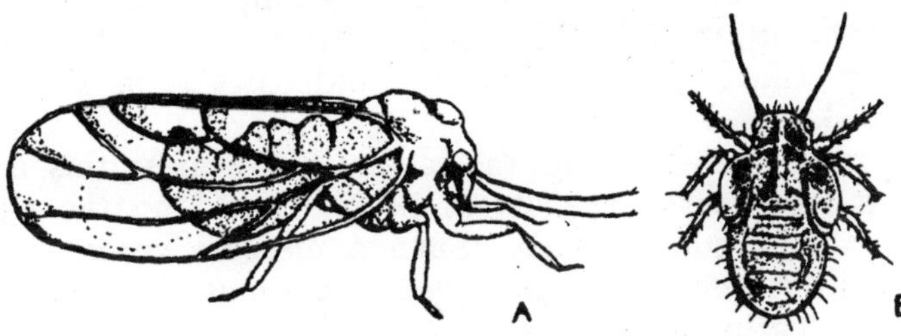

FIG. 43.—Apple Sucker (*Psylla mali*), *A*. adult; *B*. nymph.

hook-like projection at one end. They are creamy or yellowish in colour and are laid mainly on the fruit spurs and around the bases of buds, but seldom or never on old wood. With the development of the buds in April or even earlier, the eggs hatch into tiny flat, yellowish insects, which make their way into the developing leaf and blossom trusses (Fig. 43). Here they puncture the young green growth, at the same time exuding oily drops and threads which tend to prevent the normal unfolding of the bud. Gradually, as growth proceeds, the nymphs assume a green colour and the wing buds become larger with each moult. In a month or six weeks they have become fully grown, but further injury to the tree is not very apparent. The adults live until the following autumn when they lay their eggs.

Plants attacked and damage done.—The apple only is

attacked by this species, and though the one occurring on pear is very similar it seems to be a distinct species. The main injury is caused by the feeding of the nymphs on the very young leaves and developing blossom. Growth is frequently stunted and the flower-buds and their stalks may be so weakened that they shrivel and die. Sometimes they drop altogether, but they may remain on the tree, stuck together by the oily substance excreted by the nymphs and giving the appearance of the effect of severe late frost. Blossoms which develop beyond this stage are so weakened that fruit may not set and a serious loss of crop results.

Methods of Control.—As with some of the aphides on apple trees, the introduction of tar-oil washes has made the control of this insect a relatively simple matter. The eggs, which are laid in exposed positions on the tree, are readily killed by such washes, and infestation can be reduced almost to zero

Formerly limewashing was carried out against this and many other pests, the object being to seal up the eggs with a firm coat of lime. Such a measure needs to be carried out very thoroughly, good lime must be used and a thick coating must remain on the tree.

Contact insecticides are effective if properly applied (see Apple Capsid Bug), and a paraffin emulsion spraying after the apples have been gathered and before the adults have laid their eggs may greatly reduce the numbers.

FAM. ALEYRODIDÆ (*Aleurodidæ*).

White Flies, Snow Flies.

The members of this family are tiny insects with a 3 mm. expanse of wings. They are covered all over with a fine white mealy substance from which they derive their name. In many respects they are close relations of the Psyllidæ. Just before the adult form develops from the last nymph stage an inactive and resting period intervenes, which has been regarded as a kind of semi-pupal condition. The common species in this country is found in glass-houses and will serve as an example.

The Greenhouse White Fly (*Aleyrodes = Asterochiton vaporariorum*).

Description and Life History.—During the summer and early autumn this insect is found on a variety of glass-house plants. The eggs, which are provided with a small, hooked projection, are attached to the under side of the leaves and are often arranged in the form of an arc of a circle, one or more rows deep. From the egg hatches a tiny oval nymph which gradually loses its legs and antennæ with successive moults, until it enters on the semi-pupal condition during which the adult organs are built up (Fig. 44). Many generations may appear during the

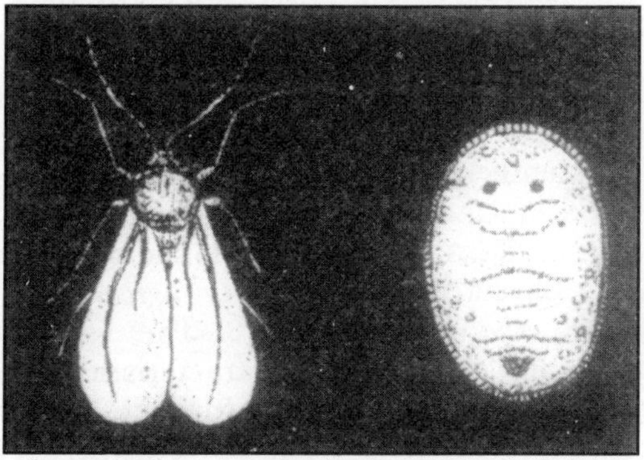

FIG. 44. White Fly (Aleyrodes) and larva. Miles.

year and parthenogenetic reproduction seems to be common.

Plants attacked and damage done.—Cucumbers and tomatoes are very commonly affected: chrysanthemums and a variety of other plants and shrubs in greenhouses and conservatories also suffer. The loss of sap causes dwarfing, and if infestation is severe the crop, in the case of cucumbers and tomatoes, is seriously affected. Moreover, once the pest has become firmly established, eradication is a tedious, costly and difficult process.

Control Measures.—Fumigation with hydrocyanic acid gas or with some other fumigant is the only successful method, and this should be resorted to before the trouble develops into serious proportions, which will rapidly be the case if the insect is neglected.

Many proprietary articles are on the market as 'White Fly fumigants.'

FAM. MEMBRACIDE and JASSIDÆ.

Froghoppers and Leaf Hoppers.

These are mostly small slender insects, which are frequently met with in summer on a variety of plants. In this country their economic significance lies in their relation to the spread of so-called 'mosaic' or 'virus' diseases of potatoes and other crops (see aphides on potatoes).

HETEROPTERA (SUB-ORDER)

FAM. CAPSIDÆ.

This family contains a large number of species, some of which are predaceous, whilst others are plant feeders. Like other members of the Rhynchota, some capsid bugs are suspected of transmitting virus diseases among plants. *Calocoris bipunctatus* is commonly found on potatoes.

The Apple Capsid Bug (*Plesiocoris rugicollis*).

Description and Life History.—The eggs, which are creamy white in colour and of characteristic shape, are laid in slits made in the bark, or in natural cracks and crevices so that they are invisible from the outside. They hatch usually late in April, and small yellowish-green nymphs emerge which have much the appearance of a small, slender aphis. They are capable of moving very swiftly and are difficult to watch, as they easily take fright and hide. On hatching they make their way to the developing flower and leaf trusses. They feed by puncturing this delicate growth, and later attack young wood and fruit. With each moult the wing-buds grow larger, and by June the bugs have become fully grown (Figs. 21 and 45). They are capable of running rapidly about the tree and have also considerable power of flight, though they appear not to migrate for very long distances and often only spread slowly throughout an orchard. Eggs are laid by the adults usually during July, and remain embedded in the bark until they hatch in the following April.

Plants attacked and damage done.—This species of capsid bug has long been known on willow and alder, but it seems to have become naturalized to the apple, on which it is now capable of carrying through its entire life-cycle. Black and red currants are also in some cases severely attacked, the leaves and young wood showing characteristic signs of the damage. The attack on blossom stalks, as in the case of apple sucker, results in so much weakening that many blossoms fall off. Leaves which are punctured exhibit characteristic curling and distortion, and around each puncture a russeted area appears. Young fruits similarly are very dwarfed and become corky in texture, whilst the points of attack on those fruits which develop become rough and pimply, the apple being generally uneven in shape (Fig. 46). Young wood is much shortened in growth and the bark cracked and rough. Leaf buds are formed very close together in consequence of the stunting, and the tree tends to assume a matted, close-growing foliage. In currants the leaf attack and the resulting crumpling and russeting are the most obvious symptoms.

FIG. 45.—Capsid Bugs: male and female (*Plesiocoris rugicollis*). Petherbridge.

Control Measures.—This insect is rapidly becoming one of the most notorious of apple pests and is recognized as being very difficult to control. Winter spraying with tar-oil washes has given variable results. In most cases some proportion of the eggs can be destroyed, but the variation in results obtained makes it impossible to say definitely that this is a means of control.

Summer spraying with a contact insecticide such as nicotine can be effective if properly carried out. The young insects are mostly sheltering among the bases of the stalks at the bottom of the blossom trusses, and the spray, to be effective, must reach them. There must, therefore, be a high pressure behind the spray and the nozzle should be one which gives coarse drops. The whole point of such spraying will be appreciated if it is realized that the operator is not concerned with spraying the tree, but with hitting the insect. A heavy, coarse, powerful spray, directed downwards into the bottom of the trusses, will achieve the

FIG. 46. - Apples damaged by Capsid Bugs.

object desired. Disturbed by the spray, many nymphs drop from the trees on to the ground, eventually ascending the tree again. If possible, trees should be grease banded before spraying or old bands renewed. As an alternative, the ground beneath the trees should be thoroughly sprayed after the tree has been done.

CHAPTER XII

NEUROPTERA—LACE-WINGS, ETC.

THIS order still contains rather diverse forms, despite the fact that many insects formerly classified as Neuroptera have been removed to other orders.

Mouthparts are of the mandibulate type.

Wings.—These are characteristically gauze-like or membranous, with a considerable number of delicate veinlets, particularly in the costal region. All four wings are similar in size and shape, and at rest are carried roof-wise over the body.

Metamorphosis.—This is invariably complex. The larvæ are carnivorous and may consequently be considered beneficial. Many of them are aquatic and show great diversity in their structure and mode of life. They usually moult twice before pupating and construct cocoons.

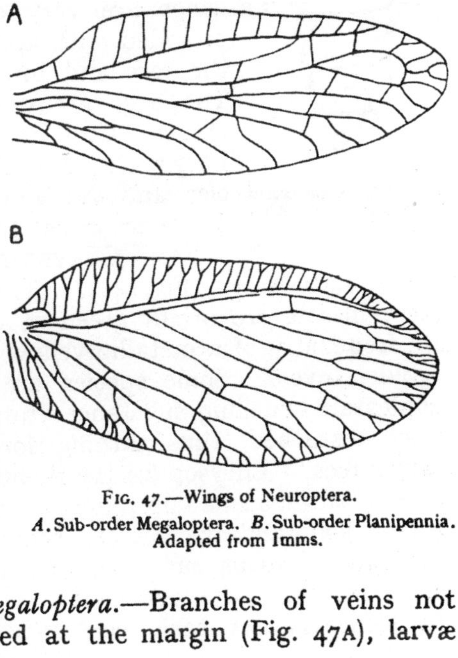

FIG. 47.—Wings of Neuroptera.
A. Sub-order Megaloptera. *B*. Sub-order Planipennia. Adapted from Imms.

There are about sixty British species of Neuroptera, distributed amongst two sub-orders.

SUB-ORDER.—I. *Megaloptera.*—Branches of veins not conspicuously bifurcated at the margin (Fig. 47A), larvæ with biting mouthparts.

SUB-ORDER.—2. *Planipennia.*—Branches of veins conspicuously bifurcated at the margin (Fig. 47B), larvæ with suctorial mouthparts.

The MEGALOPTERA includes the Alder Flies and Snake Flies which are not of economic importance.

The PLANIPENNIA, however, includes the Lace-wings, the larvæ of which do a considerable amount of good by destroying aphides, and are known as Aphis Lions (Fig. 48). The mandibles and maxillæ of these larvæ are long and somewhat sickle-shaped. They are not designed for chewing, but for seizing prey and piercing their bodies. The juices which are thus released are then sucked up along the grooves on the inner surfaces of the mandibles.

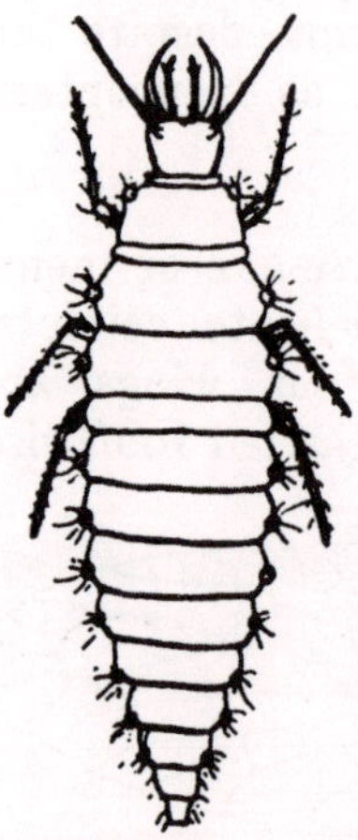

FIG. 48.—Aphis lion or Lace-wing larva. After Imms.

Individuals of two distinct families go by the name of lace-wings.

FAM. HEMEROBIIDÆ. — The so-called Brown Lace-wings belong here. The wings are very closely reticulated and provided with apparatus for interlocking the fore and hind wings of either side. The eggs are oval and have a curious minute knob-like growth upon them. Larvæ are quite smooth, free from tubercles, and usually creamy white in colour. There are about twenty British species.

FAM. CHRYSOPIDÆ. — These are the Green Lace-wings (Fig. 49). The bodies and wings of these insects are bright green in colour, whilst the eyes are frequently of a metallic yellow lustre, hence the name 'golden eyes.' Some species are able to discharge an unpleasant smelling substance from glands in the thorax when disturbed, thus earning for themselves the term 'stink flies.' Chrysopidæ lay their eggs in batches. Prior to oviposition a small drop of a sticky substance is applied to the leaf, and the female then raises the apex of her abdomen, drawing out a fine thread which at once dries and hardens. The egg is laid at the top of this stalk. In some species each egg and stalk is distinct and separate. In others the stalks are stuck together in the middle of their length, the eggs remaining free (Fig. 50).

The larvæ, which are from one-half to two-thirds of an inch

FIG. 49. - Lace-wing (*Chrysopa*) at rest.

long when full grown, crawl down the stalk on hatching out and search for aphides, thrips, psyllas and mites, which they suck dry. The empty skins are frequently cast on to the back of the larvæ which has devoured them, and, held in position by dorsal hairs, they act as a protective and camouflaging device. It has been stated that the larvæ of certain species of Chrysopidæ will devour as many as 400 aphides during its feeding period.

FAM. CONIOPTERYGIDÆ. — According to Withycombe, this family is of greater economic importance than is

FIG. 50. - Stalked Eggs of Lace-wing situated on a leaf.

commonly imagined. Members of the family are very small and fragile, somewhat resembling aphides in appearance. The larvæ are pear-shaped, tapering sharply toward the hinder end. *Conwentzia psociformis* is said to be the main check upon the Phylloxera and 'Red Spider' of the oak in this country: the cocoons of this insect may be often found in considerable numbers upon the trunks of oak trees.

CHAPTER XIII

LEPIDOPTERA—BUTTERFLIES AND MOTHS

INSECTS belonging to this order are such familiar objects that there is no need for any general description of them. The division into moths and butterflies has little scientific foundation, for all these insects possess such a remarkable similarity to one another that it is quite sufficient to separate them into families without any preliminary classification. Broadly speaking, the so-called butterflies are sunshine-loving insects and do not fly at night. Their antennæ are of a 'clavate' or knobbed type. Moths, on the other hand, are in the main nocturnal insects, and their antennæ are generally filamentous or pectinate.

Mouthparts.—By far the most prominent organ is the long suctorial trunk formed by the union of the maxillæ (Fig. 51). This organ, called the *proboscis*, is extended for the purpose of obtaining nectar from flowers, and may be seen best in the Hawk moths. When not in use, the proboscis is spirally coiled up beneath the head. The only other structures in the mouth which are well developed are the labial palps. The labium itself is very small, and mandibles as a rule are entirely wanting.

Wings.—The wings are always four in number and with few exceptions are well developed. Like the rest of the insect they are covered with overlapping scales, a character which gives the order its name. The scales furnish the brilliant colouring for which most of the Lepidoptera are noted, and which makes them very conspicuous insects. The fore wings are larger than the hind pair, and the two wings on each side are often coupled together. On the front edge of the hind wing is a bristle or a number of bristles, the frenulum, which pass through a hook or loop of hairs, the retinaculum, on the under side of the fore wing, close to its hind edge (Fig. 12). The wings are thus held more or less rigidly together. The coupling apparatus is most strongly developed in those families having a powerful flight.

Metamorphosis.—Throughout the order complex metamorphosis is the rule.

The Egg.—This stage varies greatly in its duration. Frequently the insect overwinters in this form, and the shell in this case is very hard. Eggs such as those laid by the Cabbage White butterfly hatch in the summer in a few days.

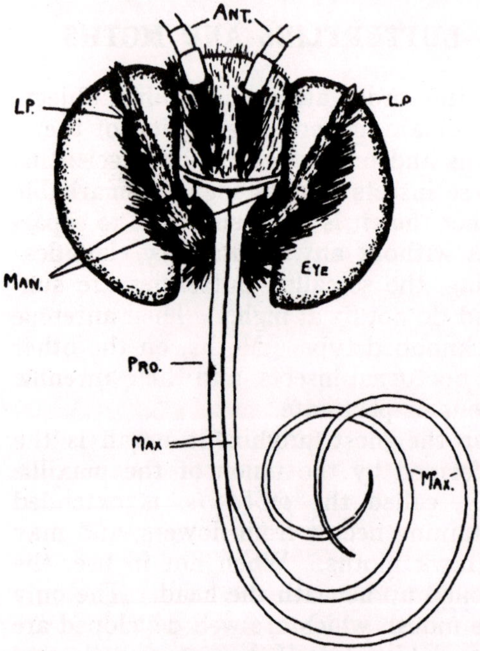

FIG. 51.—Mouthparts of Lepidopteron.
Ant. antennæ: *Lp.* labial palps: *Pro.* proboscis: *Max.* maxillæ: *Man.* mandibles.

Some Lepidoptera are noteworthy by reason of the very orderly fashion in which the eggs are laid. Those of the March Moth, for instance, are placed in a ring around a twig.

In shape a good deal of variation is also found. More or less spherical or rather flattened eggs are common, while in other species they are ovoid in shape. The shell is sculptured in various ways.

The Larva.—Most people are familiar with the caterpillar, as a Lepidopterous larva is usually called. The head is very well developed and carries mandibulate mouth parts. Behind the head are three thoracic segments each carrying a pair of jointed legs. The abdominal segments are ten in number, although the last segment but one is frequently difficult to perceive. Some of the segments are provided with a pair of pro-legs. These organs are fleshy tubercles, carrying hooks or *crochets* at the end. Abdominal segments three to six, inclusive, usually possess a pair of pro-legs each, and on the last abdominal segment another pair is present. These are somewhat modified in shape, and are known as 'anal claspers.' There are, therefore, normally eight pairs of legs altogether, and this number is never exceeded (Fig. 23, *A*). In some families fewer legs are present, the reduction always taking place in the number of pro-legs: thus in the Geometridæ only the pro-legs on abdominal segments six and ten remain, and such larvæ are called 'loopers,' from their mode of progression (Fig. 23, *B*). Some other larvæ do not develop the full number of pro-legs until they are almost fully grown. A few have pro-legs only on abdominal segments five, six and ten, those on three and four never appearing: such larvæ are known as semi-loopers.

Nine pairs of spiracles are present, one pair on the prothorax and the remaining eight pairs on the first eight abdominal segments. Caterpillars are usually more or less covered with prominent hairs, and many, when handled, produce mild skin eruptions, if such hairs are broken. Birds will not, as a rule, eat these larvæ. A great number of species produce silken threads by which they are often seen suspended from trees. The larva frequently winds a silken cocoon around itself just before it pupates, and in the well-known silk worm this silk production attains its maximum development.

The Pupa.—The silken cocoon spun by the larva provides very effective protection against enemies and weather. Those larvæ which seek protection in the soil or some other situation do not produce such a cocoon. The pupa itself is remarkable in being completely covered with chitin. It is typically 'obtect' (Fig. 25, *C*), and the chitin shell splits down the back to enable the adult to emerge. It possesses little or no power of locomotion, but the abdominal segments are capable of slight movement.

The Adult.—Some of the largest insects in the world are Lepidoptera, and probably in no other order can such a variety of wonderful colours be found. On the other hand, there are many very tiny species, and the order is sometimes roughly divided into the *macro-* (large) and *micro-* (small) Lepidoptera.

The head is always well developed and carries prominent antennæ which are 'knobbed' or filiform, and sometimes pectinate in males. The compound eyes are large and prominent. Mouthparts are very much specialized, and the suctorial trunk or proboscis is the most prominent organ. Liquid food only can be absorbed.

Of the thoracic segments the mesothorax is much the largest. The prothorax is in the form of a narrow ring, and the metathorax is also small.

The wings are coupled together in the characteristic fashion already described, and are covered with overlapping scales.

In texture the body is soft and not heavily chitinized.

Classification.—It is common to divide the Lepidoptera into Butterflies (Rhopalocera) and Moths (Heterocera). This division is based on the types of antennæ, but is very unsatisfactory. Again, the Micro- and Macro-lepidoptera have been suggested as sub-orders, but this division takes no account of many similarities of structure between the two divisions. A great many other suggestions have been made with regard to the classification of the order, many of which are too involved for elementary purposes. The one adopted here is that followed by Imms in "A General Textbook of Entomology." This system is based on the venation of the wings, associated with the structure of the mouthparts.

LEPIDOPTERA (ORDER)

SUB-ORDER.—1. *Homoneura.*—The venation of the fore and hind wings is almost identical. The spiral sucking trunk or proboscis is never present. The family Hepialidæ is the only one of economic importance.

SUB-ORDER.—2. *Heteroneura.*—The venation of the fore and hind wings is very different. The number of veins in the hind wing is much reduced. The majority of the families possess a spiral sucking trunk or proboscis. All the other Lepidoptera discussed in these pages belong to this sub-order.

Generally speaking, Lepidoptera are of greater importance to gardeners and fruit growers than they are to farmers. Vast numbers of different species have been noted as attacking fruit trees at some time or another, but many are very local, or cause such slight damage at such infrequent intervals that they are not worth mentioning in a book of

FIG. 52A. - Garden Swift Moth (*Hepialus lupulinus*), ♂ above, ♀ below.

this nature. Those who desire full details of such pests should consult books like "Insect Pests of Fruit," by Theobald, and other horticultural works.

HOMONEURA (Sub-Order)

Fam. Hepialidæ.

Swift or Ghost Moths.

The Small Garden Swift Moth (*Hepialus lupulinus*).

Description and Life History.—From tip to tip of the wings this insect measures up to $1\frac{1}{2}$ inches. The prevailing colour is yellowish-brown, but a white stripe runs from the wing base of the front wings parallel to the hind edge for a short distance, and then turns obliquely across to the apex. The legs are short and the antennæ badly developed. Eggs are dropped on to the ground during flight. The larva is milky white in colour, with a chestnut-brown head, and a plate of the same colour on the dorsal surface of the prothorax. There are also small dots on the dorsal surface of

FIG. 52B. - Garden Swift Moth (*Hepialus humuli*), ♂ above, ♀ below.

the body: from each of these arises a stiff black hair. Throughout its life the larva remains underground, and pupates in a very thin cocoon. The pupa, which is reddish-brown in colour, has prominent spines on the abdominal segments, and wriggles violently when disturbed. The moths are only on the wing after dusk, and appear in early summer (Fig. 52A).

Plants attacked and damage done.—The larva is a root feeder, attacking all sorts of market garden crops, young trees and shrubs, grass, and, in particular, strawberries.

Natural Enemies.—The larvæ are to a large extent kept

under by the attack of birds, ichneumon flies (*Alomyia debellator*, and other species), and various parasitic fungi (*Cordyceps spp.*).

Control Measures.—Swift moth caterpillars are most abundant in undisturbed soil. Perennial plants should therefore not be planted in unbroken land suspected of harbouring these caterpillars until the soil has been thoroughly turned and the larvæ disturbed and destroyed. Annual crops can be grown during this period since loss of

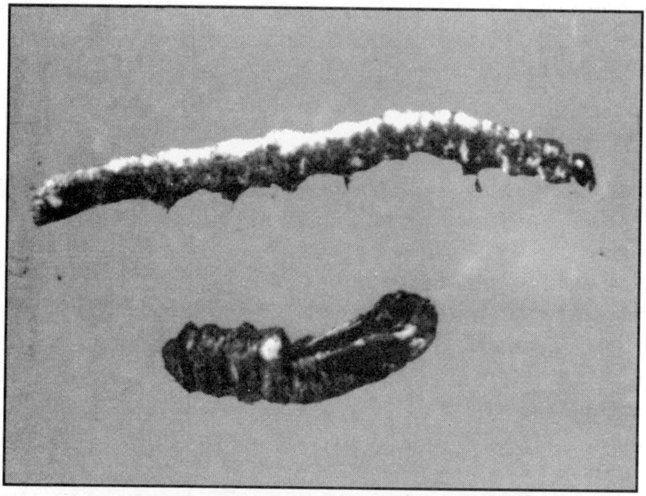

FIG. 52C. - Larva and pupa skin of *H. humuli.*

plant is not likely to be so serious as with crops intended to stand several years. If poultry can be run over the ground as it is being broken up, so much the better.

Naphthalene, at the rate of 1 oz. to the square yard, has been reported to do considerable good if well forked into the soil. So has carbon disulphide, but it is doubtful if it is advisable to use this rather dangerous and expensive chemical for a pest which can more readily be controlled by cultural methods and the use of poultry.

The Ghost Swift Moth (*H. humuli*) is a very similar species, but rather larger. It has the same life history and causes similar damage, but especially attacks hops (Figs. 52B, 52C).

HETERONEURA (Sub-Order)

Fam. Pieridæ.—The members of this family are all White Butterflies, with long, knobbed antennæ. The larvæ are typical caterpillars, which feed on a variety of agricultural and horticultural crops. The pupa is a naked chrysalis, suspended from or attached to various supports in an upright position, being fastened at the caudal end and also by a loop of silk passed round the middle region. Some species migrate in huge clouds, flying for long distances over land and sea.

The Cabbage White Butterflies.

Description and Life History.—Three different kinds of white butterflies attack cabbages and other crops, namely, the Cabbage White (*Pieris brassicæ*), the Small White (*P. rapæ*), and the Green Veined White (*P. napi*). Their habits and life histories are very similar and need not be described separately. The females lay their eggs in spring upon cabbages, either in clusters (*P. brassicæ*) or singly. The emerging caterpillars feed for about a month before pupating in yellowish white, angular chrysalides. A second brood of adults appears from these pupæ at the end of July, followed by a second generation of larvæ. The second generation frequently does much more damage than the first. Hibernation occurs in the pupal stage (Figs. 53A, 53B, 53C).

Plants attacked and damage done.—These insects confine their attentions to plants of the N.O. Cruciferæ, and do especial damage to all varieties of cabbage. They skeletonize the leaves, and still further reduce the market value of the crop by fouling it with their *frass*, or excrement.

Natural Enemies.—The caterpillars of these butterflies are heavily parasitised by various ichneumon flies, whose small yellowish cocoons can frequently be seen round dead larvæ.

Control Measures.—In small gardens hand-picking the yellowish eggs, and later the caterpillars, is undoubtedly the best way of controlling the pest. Spraying with a soap or salt solution, using $1\frac{1}{2}$ to 2 ozs. soap or salt in every gallon of water, will give good results, provided the plants are thoroughly soaked. Preparations containing Derris root have also been successfully utilized (see p. 264).

Very similar damage is caused by quite a different

FIG. 53A. - Cabbage White Butterfly (*P. Brassicae*), ♂, ♀ and underside.

insect, namely, the **Cabbage Moth** (*Barathra=Mamestra brassicæ*), of the family *Noctuidæ*.

The life history of this moth closely resembles that of the white butterflies, but pupation occurs in a shiny brown chrysalis buried in the soil. There may be two broods per year. The caterpillar is much smoother than that of the genus *Pieris*.

FAM. NOCTUIDÆ.

Cut-Worms or Surface Caterpillars.

This family comprises more species than any other family of moths. All the adults are dull coloured, at least with regard to the front wings, greys and browns predominat-

FIG. 53B. - Small White Butterfly (*P. Rapae*), ♂, ♀ and underside.

ing. The posterior wings may be brightly coloured, as in the Yellow Underwings. Noctuid moths frequently escape observation owing to the way in which the colourings blend with their surroundings. They nearly always fly at night, though species of *Plusia* may be taken on the wing in bright sunlight

The larvæ of Noctuidæ go by the name of cut-worms or

surface caterpillars. They are dull-coloured larvæ usually having five pairs of pro-legs; the larvæ of Plusiadæ, however, have four, or even only three pairs of pro-legs. They are smooth or slightly hairy, and are very difficult to identify owing to the great similarity which exists between

FIG. 53C. - Green-veined White Butterfly (*P. Napi*), ♂, ♀ and underside.

them. They remain hidden under stones, clods of earth, or in the soil during the day, and at night emerge to feed upon leaves, roots, etc.

The pupa is usually enclosed in an earthen cocoon below ground: in *Plusia* a cocoon is spun between leaves.

Noctuid larvæ are particularly destructive to crops in North America. The larva of *Aletia argillacea* is the **Cotton**

Worm, whilst the well-known **Cotton Boll Worm** is the larva of *Heliothis armigera*. The larvæ of the **Antler Moth** (*Chareas graminis*) may at times occur in countless thousands. In 1917 a severe infestation occurred in the Peak District, swarms of caterpillars devouring everything green which stood in their way. The army of larvæ swept onwards over great tracts of country until they eventually disappeared.

Life History.—The adult insects which give rise to cut-worms appear during June and July. The ribbed or reticulated eggs are laid both on the leaves and stems of cultivated and wild plants, hatching out in ten or fourteen days into the cut-worm larvæ. These feed upon any vegetable matter they come across, but of course those which attack farm plants become most noticeable. Some of the larvæ pupate in July or August, and give rise to a further brood of moths in September. From this second brood a further series of cut-worms may arise, which feed during open weather throughout autumn and winter. They pupate early in spring. Many of the summer brood, however, continue to feed right through from the time of hatching until February or March.

The most common species are the **Turnip Moth** (*Agrotis segetum*), **Heart and Dart Moth** (*A. exclamationis*), and the **Yellow Underwing** (*Triphæna pronuba*). The larvæ of the **Silvery Y Moth** (*Plusia gamma*) may at times be troublesome (Fig. 54).

Plants attacked and damage done.—Cereals, potatoes, mangolds, and turnips are the chief crops attacked. Potato tubers are hollowed out, whilst root crops may be entirely destroyed if attacked in the young state. Larger plants may have their roots hollowed out and their leaves destroyed. Cut-worms also have the habit of biting through roots just below the surface of the ground. The damage is done chiefly at night, as cut-worms avoid sunlight. The only time of year when these surface caterpillars are not active is from March until June.

Control Measures.—Since the female moths always lay their eggs on plants, land which is free from weeds has a better chance of escaping than dirty land. Similarly, if

FIG. 54. - *Typical Noctuid Moths.*

Top left : *Agrotis segelum*, male (above) and fěmale
Top right : *A. Excamationis* male (above) and Female,
Centre : *Triphoena pronuba*.
Bottom left : *Plusia gamma*, *Hadena oleracea*.
Bottom right : *Mamestra brassicae*, *Mamestra pericariae*.

grass land intended for wheat is broken up early in July, there is less chance of cut-worm attack, following the deposition of eggs. The use of poison baits has proved very successful in North America, and experiments in this country have also given good results. It is probable that the use of poisoned bran after the manner suggested for leather-jackets would be most useful in coping with bad attacks of cut-worm in root crops (see p. 200).

Poultry readily devour cut-worms, and it is an excellent plan to allow poultry to forage over land which is known to contain these pests.

FAM. GEOMETRIDÆ.

'Loopers.'

The larva is very characteristic in this family, and is of the well-known 'looper' type. Larvæ of Geometers usually have pro-legs upon the sixth and tenth abdominal segments only. In progressing from one spot to another the caterpillar arches its body until the pro-legs on the sixth abdominal segment are close to the posterior pair of thoracic legs, and then it extends the forepart of the body (Fig. 23). At rest the larva usually extends itself rigidly with its fore part in the air, holding on to twigs, etc., by its pro-legs. Owing to this habit, and to their protective colourings, looper larvæ frequently escape observation, since they very closely resemble their surroundings.

The pupa has the first two pairs of legs rather longer than is usual amongst Lepidoptera. Some pupæ are enclosed in a slight silken cocoon, whilst others are buried in soil.

The adult is always a slender individual, whose wings may show great differences as between the sexes. Frequently the male may have relatively large wings, whilst the female is almost, if not entirely, apterous. At rest the wings are extended horizontally. Flight usually takes place at night.

A number of species are at times very destructive to fruit trees and bushes: the most important are described below.

The Winter Moth (*Cheimatobia brumata*).

Description and Life History.—The male Winter Moth has a wing expanse of from 1 to 1½ inches. The wings are fringed with hairs: the fore wings are brownish-grey with distinct, darker wavy lines across them, whilst the hind wings are grey without the markings (Fig. 55A). The female has a much shorter, thicker body, with long thin legs and very tiny withered-looking wings (Fig. 55B). She is quite incapable of flight. The adults emerge about the middle of October from pupæ which are slightly buried in the soil. The female at once crawls up the stems of fruit trees and along the branches as far as the terminal shoots. Here she deposits her small, cylindrical, light green eggs, laying them usually in clusters at the bases of the buds. Here they remain until the following March, turning to a reddish colour as they get older. Just before the buds open in spring the eggs hatch into small grey 'loopers,' which eventually become green in colour, with paler stripes along the body. The caterpillars feed voraciously upon leaves, buds and blossom: they frequently spin together the edges of leaves with silk threads, and live under this protection. In May and June they lower themselves down from the trees into the soil, where pupation occurs in a cocoon made of particles of soil.

FIG. 55A. - Winter Moth : male Natural size.

FIG. 55B. - Winter Moth : female. Natural size.

Plants attacked and damage done. — About the only kinds of trees which are not attacked by Winter Moth caterpillars are the conifers. All kinds of fruit trees may be severely attacked, particularly apples and pears. Very considerable damage may be caused in bad cases, the trees being entirely stripped of their foliage by the beginning of June, and presenting a bare, winter-like appearance. Buds, leaves, blossom and even young fruitlets may be devoured. In mild cases the crop of fruit is diminished: in severe cases the entire crop is lost, whilst the trees are seriously weakened through having to put forth a fresh crop of leaves during midsummer.

Natural Enemies.—Birds prey upon the caterpillars, which are also parasitised to a slight extent by hymenopterous parasites.

Control Measures.—The case of the Winter Moth shows very clearly how necessary it is to have a working knowledge of the life history of a pest before control measures can be devised. Knowing that the female Winter Moth is

FIG. 56. - Fruit tree with grease band.

incapable of flight and must therefore crawl up the trunk and branches of fruit trees in order to lay her eggs, it is at once possible to devise preventive measures. These take the form of 'grease banding,' which is nowadays a routine operation in all well-managed orchards. The band consists firstly of a strip of waterproof paper, about 8 inches wide. This is firmly tied round the tree at a height of from 3 to 5 feet, selecting a smooth piece of bark so that the insects may not crawl underneath in crevices, and get past (Fig. 56). On the paper is then smeared a complete ring, not less than 3 inches wide, of one of the grease-banding compositions now on the market. Those of a sticky nature are to be preferred to those which are oily, since they are more efficient and last longer. The banding should be begun in the first week of October, and the bands should be kept tacky until the following April. Bands which are placed low down are not so efficient as those at a height of 3 to 5 feet: it may be necessary, therefore, to put separate bands upon each of the main branches of some dwarf bush fruit.

Trees which in spring show signs of infestation by Winter

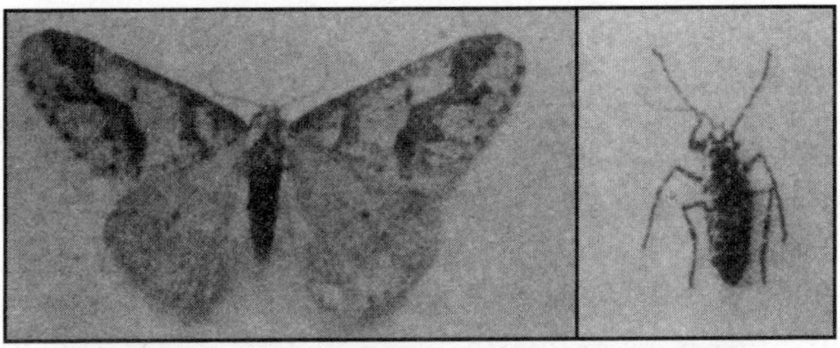

FIG. 57A. - Mottled Umber Moth, male and female.

Moth caterpillars should be sprayed with a stomach poison, such as arsenate of lead (see p. 262).

Two closely related species do very similar damage to that caused by the Winter Moth.

The Mottled Umber Moth (*Hybernia defoliaria*) is nearly twice as large as the Winter Moth and is browner in colour. The female is quite devoid of wings (Fig. 57A). The larva is usually chestnut brown with dark wavy lines along the back. The moths appear over the same period as the Winter Moth, namely, October to January.

The March Moth (*Anisopteryx æscularia*) appears in the month of March (Fig. 57B). The female is quite wingless and lays its eggs in bands round the smaller twigs after the style of the Lackey Moth, though the eggs are much smaller. The larva is very similar to that of the Winter Moth.

The preventive and control measures previously described are sufficient to keep these pests in check.

The Magpie Moth (*Abraxas grossulariata*).

Description and Life History.—The Magpie, or Currant and Gooseberry Moth, is too well known to need detailed description. Usually it is almost two inches wide across

FIG. 57B. - March Moth, male and female.

the wings: these are creamy white in colour, with characteristic spots and bands of black and yellow: there are black spots on the dorsal surface of the body (Fig. 58). Adults appear during July and August, the female laying her oblong, yellowish eggs upon the leaves of currants and gooseberries. In from ten to fourteen days the eggs hatch into tiny black caterpillars. These feed upon the foliage of the bushes, but do very little harm during the season in which they hatch out. They desert the bushes towards autumn, and retire into whatever shelter they can find. They may frequently be found under fallen leaves round currant bushes, in crevices in walls and fences, etc. Here they remain, still in the larval state, until the following spring. When the fruit bushes start putting forth leaves in spring, the caterpillars emerge and feed upon them, moulting until they reach a length of from 1 inch to 1¼ inches. They are then black and white in colour, with yellow or orange markings at the side (Fig. 59). The larva pupates in June in a shiny black chrysalis, with yellow bands round

FIG. 58. - Magpie Moth (*Abrazas grossulariata*). Natural size.

the abdominal segments. Usually the pupa is enclosed in a delicate silken cocoon fastened to the under side of leaves, to walls, fences and twigs. The adult emerges in about a month. There is one brood only per year as a rule.

Plants attacked and damage done.—Currants and gooseberries suffer most damage, but the pest also attacks the plum, sloe, apricot, spindle tree, and nuts. Damage is done through the destruction of the leaves, with consequent reduction or total loss of crop and weakening of the bushes. It should be noted that the damage is done early in the year, as the leaves are appearing. The amount of damage caused by the newly hatched autumn larvæ is insignificant.

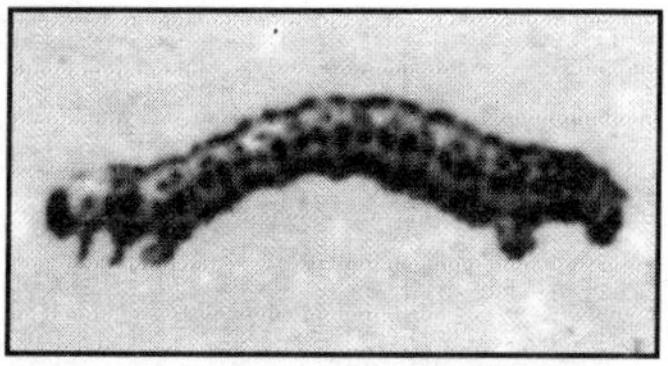

FIG. 59. - Larva of Magpie Moth.

Natural Enemies. — Birds search for and devour the hibernating larvæ.

Control Measures.—The destruction of winter shelter, so far as it is practicable, assists in preventing attacks. In this connection it may be noted that Magpie Moth damage is most common in small, enclosed gardens, where there is an abundance of shelter during winter: large, exposed plantations are less subject to attack.

Spraying with a stomach poison is necessary to kill caterpillars which swarm over the bushes, but if arsenical preparations are employed they must be used as early as possible (*i.e.* before flowering begins) to avoid contaminating the fruit. If this early spraying cannot be carried out, hellebore powder should be substituted for the arsenical compounds, as this may be used within a month or three weeks of the date when fruit-picking is to begin. The following sprays have been recommended by the Ministry of Agriculture:

(*a*)	Lead Arsenate paste	½ lb.
	Water	10 gallons.
(*b*)	Hellebore powder	2 lbs.
	Soft soap	½–1 lb.
	Water	10 gallons.

FAM. TORTRICIDÆ.

Tortrix Moths.

This is a large family containing a number of destructive pests. The adults are usually small in body, but proportionately large as to the wings: they fly at dusk. The larva has five pairs of pro-legs and a distinct, brownish head. When disturbed, tortrix larvæ usually wriggle themselves backward with extraordinary energy. They remain concealed in fruit or stems during the feeding period, or spin together leaves for shelter. Another characteristic of these caterpillars is their habit of letting themselves down from leaves and twigs by means of a silk thread. This is their normal method of reaching the ground: the larvæ also frequently drop from their supports and remain suspended when interfered with. *Tortrix viridana* is a species which frequently defoliates the oak.

FIG. 60. -Codling Moth (*Cydia pomonella*).

The Codling Moth (*Cydia pomonella = Carpocapsa pomonella*).

Description and Life History.—The adult moth is about half an inch wide across the wings, which are greyish with a silky sheen. There are darker patches at the base and tip of each of the fore wings, whilst the male has in addition a dark line on the under surface (Fig. 60). The female is in flight during the latter part of April and May, but owing to its habit of remaining motionless on tree trunks during the day it is not often seen. The female lays her small, glistening eggs singly on the tiny, developing fruit of apples and pears, and sometimes on leaves or twigs. Larvæ hatch out in about ten days. These at once crawl over the young fruit in search of the 'eye,' or depression at the distal end of the apple where the remains of the calyx and other parts of the flower are situated. The larva bites its way through the eye into the interior, and feeds upon the core and the surrounding tissue (Fig. 61). In less than a month the caterpillar is full fed, and is then about half an inch long, creamy white or pink in colour, with a brown head. It burrows a hole in the side of the apple and, according to Theobald, if the fruit is still on the tree, it either lowers itself to the ground by a thread or crawls down the trunk of the tree. Generally, however, the affected fruit drops before the larva is full fed. The larva then spins itself a silken cocoon in any convenient shelter, more especially in crevices in the bark of fruit trees, wherein it lives as a larva until the following spring. Pupation

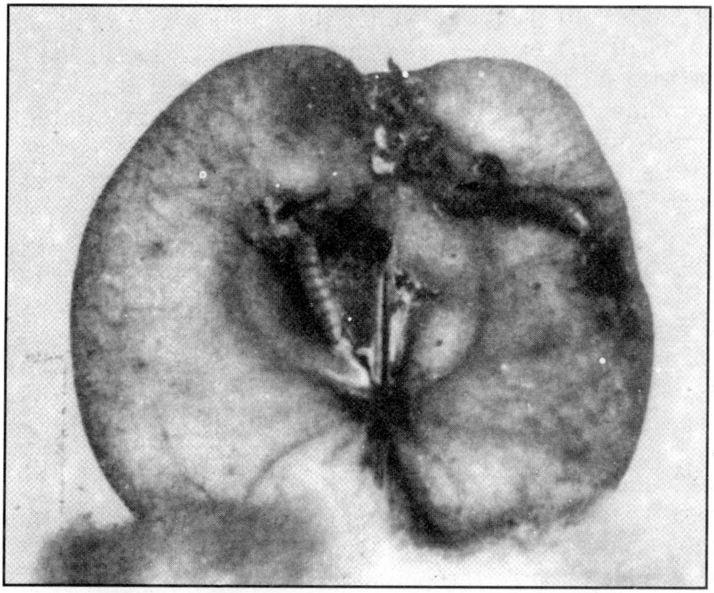

FIG. 61. -Apple infested with larvae of codling Moth.

takes place in spring within the larval cocoon. In this country there is normally only one brood per season : in America there may be two or more broods.

Trees attacked and damage done.—All varieties of apple are very liable to attack by 'maggot' : pears and plums are also attacked, but to a lesser degree. Attacked fruit nearly always falls off the tree before it is ripe, and if it does hang on long enough to be picked, its market value is greatly reduced, if not entirely destroyed. The centre of the apple becomes churned into a brownish mass of larval excrement, or frass, whilst the 'maggots' themselves frequently remain inside, with unpleasant results.

Natural Enemies.—Tits in particular devour the hibernating larvæ. The larva itself is parasitised by hymenopterous insects, but not to a degree sufficient to keep it in check.

Control Measures.—Old trees with much rough bark upon them are very liable to attacks by this pest, since the caterpillars find plenty of winter shelter close at hand. A clean, healthy tree, free from moss and loose bark, offers very little shelter for wintering larvæ. Apple trees, therefore, cannot fail to benefit from a spraying in winter, designed to clean the trunk and branches. If large quantities of apples have fallen they should as far as possible be destroyed before the caterpillars escape: pigs make good scavengers. On a small scale the larvæ may be destroyed by enticing them to form their cocoons in bands of sacking tied round the tree trunks: these must be burnt in autumn.

To destroy maggots already upon the fruit, a stomach poison such as lead arsenate must be employed. Spraying for leaf and fruit-eating insects must be performed either just before the blossom has opened, whilst it is in the 'pink' stage, or just after the fruit has set, and the discoloured petals are falling. The later spraying is specially suitable against the Codling larva, for with a fine nozzle it is possible to force the poisonous fluid into the still open 'eye' of the young apple, where a deposit of arsenic is formed. When the caterpillar seeks to enter through this region, its first meal proves fatal. Spraying should not be carried out until the petals begin to drop, owing to the danger of poisoning bees.

During 1925–1926 there was a number of cases of severe illness and even death, alleged to have been due to the consumption of apples containing arsenic. In all cases the suspected apples proved to be of foreign origin. It has been definitely shown that there is absolutely no danger of arsenic poisoning from English-grown apples, where spraying is carried out at the normal times. In some experiments carried out at Harper Adams Agricultural College in

1926, Newton Wonder apple trees were sprayed with lead arsenate fluid on 24th April and 31st May. Apples gathered on 8th July contained only .0013 grain of arsenic per lb. of apples, whilst those picked on 5th August contained only .00028 grain per lb. Unsprayed apples were found to contain about .00028 grain arsenic per lb. as part of their normal composition, and since the law limits the amount of arsenic present in food-stuffs to .01 grain per lb. of food, it is clear that the amounts mentioned above are well within the legal limits and unlikely to cause poisoning, especially when the earliness of the picking is borne in mind.

In America, where there may be two or three broods of Codling Moth per season, spraying with poisons has to be carried on until quite late in the season. The climate in the fruit-growing districts is also much drier than it is here, so that there is less chance of the poisons being washed off the fruit. Consequently there is more danger of arsenic-poisoning from these apples than from home-grown ones. The largest proportion of the arsenic is found in the regions of the eye and stalk. The danger of arsenic poisoning from eating apples has been greatly exaggerated; it is understood that American fruit growers are so modifying their spraying arrangements as still further to reduce the already quite slight possibility of arsenic reaching the consumer.

The Pea Moth (*Cydia nigricana*) is the cause of 'maggoty' peas. The adult moth is about half an inch wide across the slaty grey wings, and it lays its eggs during June in the blossom of peas. The small yellowish larva enters the pod, and later burrows into the developing peas when these are of a fair size. When full fed the larva finds its way into the soil and there pupates. Early peas generally escape, and peas grown in the field are not attacked so frequently as garden peas.

The pest is exceedingly common, and causes much damage in bad seasons. A certain check can be imposed upon it by burning the haulm after an attack, thereby destroying any larvæ which still remain in the pods. Any method of destroying the cocoons in the surface soil, such as exposing them to birds, or burying them by deep cultivation, will also be beneficial.

FAM. BOMBYCIDÆ.—These moths are very hairy and heavily built. The larvæ are hairy, and pupation always occurs in a dense silken cocoon. The larva of *Bombyx mori* is the Chinese silkworm.

The Lackey Moth (*Malacosoma = Clissiocampa neustria*).

Description and Life History.—The male of the species is about 1 inch and the female 1½ inches across the wings. The colouring is very variable, usually reddish brown to red as to the fore wings, with two light bars across them (Fig. 62). They fly only at night. Eggs are laid in bands round small twigs in July and August: there may be from 50 to 200 of these greyish, slightly hollowed ova in each band (Fig. 63). The eggs remain dormant until the end of the following April, when they give rise to larvæ, which when mature are brilliantly striped from head to tail with lines of blue, orange-red, white and black. Lackey Moth caterpillars live in tents or webs of silk which they construct amongst the leaves and branches of the tree they favour. They feed upon leaves, and as they grow forsake their nests and spread over the trees. When full fed they spin loose cocoons amongst leaves or upon fences, and pupate. From the enclosed, brownish chrysalis, there emerges the adult in from two to three weeks.

FIG. 62. -Female (above) and Male Lackey Moths. Natural size.

Plants attacked and damage done.—The apple and pear are most frequently attacked, but cherry, plum, oak, rose, hawthorn and many other plants also suffer. Damage consists in the defoliation of the tree and consequent loss of crop. Plagues of Lackey Moth caterpillars occur at intervals, the intervening seasons being more or less free from them. The pest is much less common in the north than in the south.

Natural Enemies.—At least sixteen kinds of ichneumon flies and two species of braconid fly parasitise the eggs and larvæ of the Lackey Moth. Birds do not seem to care much for them.

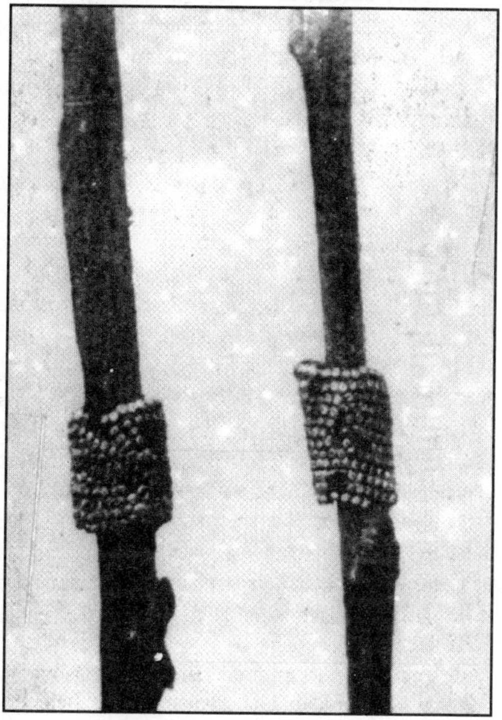

FIG. 63. -Egg bands of Lackey Moth.

Control Measures.—Early spraying with lead arsenate is effective if carried out before the nests or webs become too large or numerous. On a small scale much good may be done by cutting out the nests and burning them. The egg bands ought to be destroyed whenever noticed during pruning operations.

The Vapourer Moth (*Orgyia antiqua*) at times causes damage to the foliage of fruit and forest trees and also to herbaceous plants. The male has a wing expanse of just over an inch : the wings are chestnut brown with a white, crescent-shaped spot towards the hind angle of each fore wing. The female has a short, stout body but only the vestiges of wings. She lays eggs in large greyish masses upon her empty cocoon or upon surrounding leaves, but never moves far from her cocoon. The emerging caterpillars are very hairy with four tufts of hair along the back. Pupation takes place in a yellowish grey cocoon. There may be more than one brood per year.

Control Measures.—Egg masses may be destroyed in winter: lead arsenate or other stomach poisons account for the larvæ, which usually are not severe pests.

FAM. COSSIDÆ.—These insects are amongst the largest of British Lepidoptera, with big hairy bodies and wings. They are nocturnal in habit and lay their eggs upon the bark of trees. The larvæ are provided with very strong mandibles and they bore out and live in tunnels in the wood, frequently causing severe damage.

The Goat Moth (*Cossus cossus = C. ligniperda*).

Description and Life History.—The body of the female is 1½ inches long, and the wing expanse over 3 inches. Wings are mottled pale brown and grey, with dark streaks across them. The body is very broad and hairy with a number of greyish bands round it. The females are on the wing in June and July, laying eggs in groups of from ten to fifty, deep in crevices of the bark of various trees. The larva at first is pink, but later it turns to a yellowish red, with two brown spots on the first segment. It is over 3 inches long when full grown. It lives in tunnels in the wood, remaining in the larval state for two or three years, and giving off a peculiar goat-like smell. Pupation occurs in a chrysalis near the outside of the tree, the pupa lying covered with silk close to the opening of the burrow.

Trees attacked and damage done.—Forest trees are usually most favoured by this pest, but apples and pears are occasionally damaged. The attack may extend over a period of years, and attacked trees usually die.

Control Measures.—There is no easy method of controlling this pest. Badly attacked trees will undoubtedly perish, so that it is advisable to cut them down and have them cut up for burning so as to destroy the larvæ. Valuable trees may be saved in the early stages of attack by injecting small quantities of carbon disulphide into the larval galleries and sealing up the exits with clay. There are no practicable preventive measures against attack.

The Wood Leopard Moth (*Zeuzera æsculi*) has a similar life history and does similar damage.

FAM. SESIIDÆ (ÆGERIIDÆ).

Clearwings.

The species belonging to this family are characterized by the absence of scales from the greater part of the wings, which therefore seem transparent. The fore wings are very narrow, and the body is slender. There is frequently a curious fan-like expansion of scales at the posterior end of the abdomen (Fig. 64). These moths are diurnal, and owing

FIG. 64. -A typical Clearwing Moth. (*Sesia* sp.)

to their bright colourings often resemble bees, wasps and ichneumons. The larvæ and pupæ live in the wood of trees and shrubs.

The Currant Clearwing (*Sesia*=*Ægeria tipuliformis*).

Description and Life History.—The imago is about two-thirds of an inch across the wings, which are transparent, with edgings of black and orange. The body is shining black, with three yellow stripes across it in the female, and four in the male. They appear in June and lay eggs singly on the stem of currants, near the buds. The larva is creamy white, with black spots on each segment, and a shiny brown head. It bores into the pith of the shoot and pupates there the following May, after feeding throughout the winter on the pith. The chrysalis forces its way partially out of the stem by means of its spines, using the hole made by the caterpillar on entering.

Plants attacked and damage done.—Red, white and black currants are affected. The shoots wither in spring and the leaves and fruit may be seriously stunted. The pest is very local, but on occasion may cause very serious losses.

Control Measures.—These consist in pruning back infested shoots in January and February, and again in the summer, until no signs of tunnelling by the larva can be seen. The prunings should be burnt.

TINEINA VERA

Into this group have been placed over 700 British species of moths, all of which are small and whose relationships are uncertain. A number are injurious to cultivated plants. All individuals belonging to the group are small, narrow-winged insects, with long hair fringes to the wings, but the classification of the group is as yet very indefinite.

FAM. PLUTELLIDÆ.

The Diamond Back Moth (*Plutella maculipennis* and *P. cruciferarum*).

Description and Life History.—The adult is only a quarter of an inch long. The fore wings are grey, with three whitish, triangular markings on the posterior margin. When at rest the insect carries its wings roof-wise over its back, so that the triangular patches coincide and form diamond-shaped markings. The hind wings have a fringe of very long hairs. Eggs are laid during June and July upon cruciferous plants, such as turnips, charlock, etc. The grey-green larvæ feed on the underside of the leaves for about a month, spinning a slight web. They pupate in a silk cocoon, those of the second brood remaining over winter. There may be two or more broods per year (Fig. 65).

Plants attacked and damage done.—Only cruciferous plants are damaged to any extent. Swedes, turnips and kale suffer the most, but rape and mustard are also attacked. In feeding, the larva eats away the lower epidermis of the leaf, together with the mesophyll (or middle portion), but leaves the upper epidermis and the main veins untouched. The pest is very sporadic in appearance. For many years there may be no reports of damage due to this insect, and then there may be a sudden plague causing thousands of pounds' worth of damage. The moth occurs most frequently along the east coast of Northern England and Scotland: it caused considerable damage to swedes in parts of Norfolk and Cambridgeshire during the summer of 1928.

Natural Enemies.—Ichneumon flies of various kinds

FIG. 65. -Diamond Back Moth (*Plutella maculipennis*) much enlarged.

parasitise the larvæ. Lapwings and starlings also devour them in enormous numbers. Wet, stormy weather is fatal to the larvæ, which thrive best under warm, dry conditions.

Control Measures.—Owing to the intermittent nature of the attacks of this pest, little positive information is available as to successful control measures. The problem is made more difficult because the caterpillars feed on the underside of leaves. Knocking off the larvæ with twiggy branches fixed to scufflers has been successfully tried. A powder spray, consisting of lime and soot mixed with nicotine powder, would probably give the best results, especially if some device for turning up the leaves were attached to the sprayer.

Other moths belonging to the *Tineina vera* worthy of mention include the following :—

The Raspberry Moth (*Incurvaria = Lampronia rubiella*).

Description and Life History.—This purplish-brown moth is less than half an inch across the wings. It appears early in June, laying eggs in the flowers of the raspberry. The larvæ live for a short time in the fruit, but at this stage do little harm. They drop to the ground and hibernate as larvæ in any available shelter, forming a slight cocoon. The following March and April the young caterpillars crawl up

FIG. 66.-Raspberry Moth (*Incurvaria rubiella*) enlarged.

the canes and burrow into the shoot via the buds. Here they turn red, having a black head. Pupation takes place within the shoot, which eventually dies.

Nature of Damage.—Young shoots die off above the point of entry of the larva, and thus there is considerable loss of crop.

Control Measures.—Old canes should be cut back to the ground in autumn and at once burnt. Old stakes should be removed and the soil round the canes should be forked over to expose any larvæ to the attention of birds. A spraying with lead arsenate late in March, followed by another spraying a few days later, will leave a deposit of arsenic on the buds and kill any caterpillars trying to enter. In very bad cases it may be necessary to cut back the canes in spring to within a foot of the ground. This prevents the formation of flowers, and the female moths consequently have nowhere to lay their eggs.

The Small Ermine Moths (*Hyponomeuta spp.*).

Description and Life History.—There are a number of species of small ermine moths: they are all about three-quarters of an inch across the wings. The fore wings are whitish, with black dots, and the hind wings a darker grey (Fig. 67). The moths appear in July and August, and lay their eggs, which are cemented together into scale-like masses. The young larvæ hatch out in September and remain hidden over winter under the egg shells. The following spring they commence to feed on the buds, and later on the leaves. In June large webs or tents are formed in which live colonies of almost black caterpillars. Pupation takes place in June in whitish, spindle-shaped cocoons arranged in nests on the under sides of leaves: these take about a fortnight to develop into the adults.

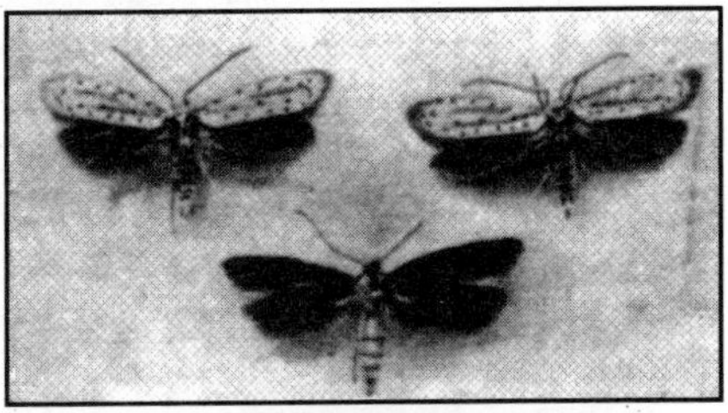

FIG. 67. -Small Ermine Moths. Natural size.

Plants attacked and damage done.—Apple trees are most frequently attacked, but pears, plums, cherry, hawthorn and mountain ash also suffer. Damage is caused by defoliation of the tree.

Natural Enemies.—Several species of ichneumon flies parasitise the larvæ of Ermine Moths.

Control Measures.—It is probable that the young larvæ over-wintering beneath the egg shells would be killed by a winter wash composed of one of the tar-oil compounds. A lead arsenate spray effectually deals with the larvæ if applied before the tents begin to form.

CHAPTER XIV

COLEOPTERA—BEETLES

THE Beetles constitute the largest order of insects. Nearly 200,000 different species are known, and of these over 3000 are found in the British Isles. In habits they are more varied than most other insects, some being purely aquatic and some terrestrial. Many are incapable of flight, while others are very strong on the wing.

Mouthparts.—Beetles possess typical mandibulate mouthparts. The mandibles themselves are often extremely large, and in the males of some species (Stag Beetle) are so unwieldy as to be useless for feeding purposes.

Wings.—The structure of the wings is very definite throughout this order, no other insects having become so specialized. Two pairs of wings are present, one pair on the mesothorax and one pair on the metathorax. Those on the mesothorax are hardly recognizable as wings, being very hard in texture, due to chitin laid down on them. They are known as *elytra* (Sing. *elytron*). When the insect is at rest, these elytra cover the dorsal surface, meeting in the median straight line, and form an efficient means of protection. One sub-division of the order contains insects in which these elytra are very short and do not cover the abdomen (Staphylinidæ). Elytra are not used in flight. The metathoracic wings are very large. They are hinged in such a way that the tip can be folded back upon the rest of the wing, so that the whole can be covered by the elytra. These metathoracic wings are the only ones which support the insect in flight.

In certain beetles the power of flight has been lost. The elytra are joined along a mid-dorsal line and tightly secured to the body. Such insects are found typically among the Carabidæ (Ground Beetles) and some of the Rhyncophora (Weevils).

Metamorphosis.—The metamorphosis is invariably complex. Definite larval and pupal stages are present, but the larvæ exhibit many different forms.

The Egg.—In the majority of cases the eggs are roughly oval in shape and of comparatively short duration. They are thin shelled and are usually concealed.

The Larva.—Two well-defined types are found among beetle larvæ. All are alike in possessing a definite head with mandibulate mouth-parts, but legs may or may not be present. Legless larvæ are common to the weevils and many other beetles, especially wood-borers. When legs are present they are found only on the thoracic segments, and there are always three pairs. Pro-legs such as are seen in Lepidopterous larvæ are never present, but the terminal segment of the abdomen sometimes carries an anal sucker or pseudopod, and a pair of cerci. There are ten segments in the abdomen.

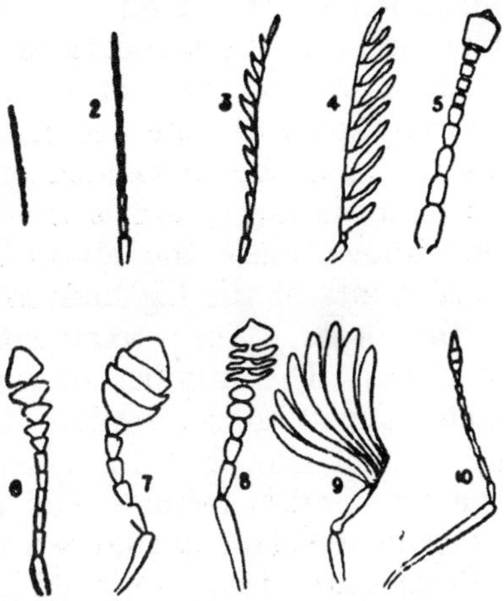

FIG. 68.—Types of Antennæ belonging to Coleoptera.

1. moniliform (beaded). 2. filiform (thread-like). 3. serrate (saw-like). 4. pectinate (comb-like). 5. capitate (with a head). 6 and 7. clavate (clubbed). 8. perfoliate-clavate. 9. lamellate (plated). 10. geniculate (elbowed). From Hall's "Common British Beetles."

The spiracles are situated in a row down each side of the larva. They are usually well marked and plainly visible.

The Pupa.—The pupa is usually of the free or libera type. It is not adapted for withstanding extreme climatic conditions, and usually gives rise to the adult in a short time; two or three weeks is the usual length of the pupal instar.

There is no silken cocoon surrounding it, but those larvæ which pupate in the soil construct an oval cell, in which the pupa is formed. Some larvæ, such as those of the Coccinellidæ, pupate on bushes where they have been feeding on aphis colonies. Such pupæ are exposed to weather and in consequence are partially chitinized. This type is the *semi-obtect* pupa.

The Adult.—The whole body of an adult beetle is covered with well-developed chitin plates, except the dorsal surface of the abdomen, which is protected by the elytra.

The head is a very hard capsule, and the mouthparts are always of a mandibulate type. In the weevils the mouthparts are situated at the tip of a long rostrum, which is an elongation of the head. Antennæ vary very greatly in structure, and the sub-divisions of the order are to a large extent based on these differences (Fig. 68). Four wings are normally present, the mesothoracic pair being hardened (elytra), and the metathoracic pair are very large. The prothorax is large, and is the only thoracic segment seen from above, since the elytra cover the remainder. The tarsal joints of the leg form another basis of classification of the order. They vary considerably in number and structure. The number of segments in the abdomen is often difficult to determine. Usually five can be seen on the ventral side, but there may be more. The terminal segment is often tubular, and acts as an ovipositor which is capable of being withdrawn into the body.

Beetles are often long lived in the adult stage, and many adults feed voraciously. For this reason they are often responsible for damage to plants to an even greater extent than their larvæ.

Classification.—Many systems have been proposed, but so far none are very satisfactory, since a great many exceptional forms exist.

The following classification is adopted here, since from the elementary point of view it does not present great difficulties :

COLEOPTERA (ORDER)

- *Adephaga* (SUB-ORDER)
- *Polyphaga* (SUB-ORDER)
  - SUPER-FAM.: **Staphylinoidea**, **Diversicornia**, **Heteromera**, **Phytophaga**, **Rhyncophora**, **Lamellicornia.**

SUB-ORDER.—1. *Adephaga.*—Antennæ generally filiform, tarsi five-jointed. Larvæ with three pairs of legs, active and predaceous.

SUB-ORDER.—2. *Polyphaga.*—Very diverse types of antennæ and tarsi. Larvæ with or without legs.

SUPER-FAMILIES.—1. *Staphylinoidea.*—Generally simple or clavate antennæ, tarsal joints variable. Elytra often short and not covering the abdomen. Larvæ with three pairs of legs, active, and some predaceous (p. 134).

2. *Diversicornia.*—Antennæ and tarsi very variable. Larvæ with or without legs (p. 136).

3. *Heteromera.*—Antennæ variable. Legs on pro- and meso-thoracic segments have five-jointed tarsi, meta-thoracic legs have four-jointed tarsi. Larvæ with three pairs of legs (p. 147).

4. *Phytophaga.*—Antennæ generally simple, tarsi apparently four-jointed. Actually the fourth joint is very small and closely united to the fifth, which conceals it. Larvæ with short or reduced legs, and in some cases legs are absent (p. 148).

5. *Rhyncophora.*—Head more or less elongated into a rostrum. Antennæ usually geniculate and clavate. Tarsi apparently four-jointed, but actually with five joints. Larvæ generally without legs (p. 156).

6. *Lamellicornia.*—Antennæ in the form of a lamellate club. Tarsi five-jointed, and legs often modified for digging purposes. Larvæ with three pair of legs, abdomen usually curled ventrally towards the head (p. 164).

ADEPHAGA (SUB-ORDER)

FAM. CARABIDÆ (**Ground Beetles**) and CICINDELIDÆ (**Tiger Beetles**).

Description and Life History.—Ground Beetles are very common insects, and the majority are black or dark coloured. They are frequently found whenever soil is being dug, and often shelter under heaps of refuse, sacks, boards, etc., lying on the ground. The majority are carnivorous in habit, and are capable of running rapidly. Legs are well developed for this purpose, and so fixed has this kind of locomotion become that the wings are not used. The elytra are usually firmly fixed down upon the back so that the insects cannot fly. Little is known of the duration of the various stages in their life history. The larvæ are very active and predaceous, having well-developed thoracic legs and powerful mandibles. The head projects forward from the thorax after the usual manner of carnivorous larvæ. A few of the Ground Beetles have been accused of damaging cultivated plants, but this type of feeding seems to be exceptional in the family. *Harpalus ruficornis* (Fig. 69) and *Nebria brevicollis* occasionally do damage to strawberries by making holes in the fruit, and *Pterostichus madidus* (Fig. 69) has been accused of 'cutting off' mangolds and sugar beet at ground level, when the roots are about half an inch in diameter.

The Tiger Beetles (Fig. 69) are bright-coloured insects, usually green, and are entirely predaceous. Their larvæ live in small burrows in the soil, where they lie in wait for passing insects which they catch and devour.

The whole of the Adephaga may be considered, in the present state of our knowledge, to possess very little economic significance.

POLYPHAGA (Sub-Order)

Super-Family.—1. *Staphylinoidea*

The beetles of this group are characterized by more or less abbreviated elytra, and some of the abdominal segments are always visible from above. The majority feed upon dead animals and vegetable matter, but a few are predaceous. A large species, *Ocypus olens* (Fig. 69), the **Devil's Coach Horse,** is very common. When disturbed it assumes a scorpion-like attitude, spreading out its legs and curling its abdomen upwards over its back. Two other small Staphylinid beetles are known to bore into and destroy the pupæ of the Cabbage Root Maggot Fly, and must therefore be accounted beneficial insects from man's point of view.

The larvæ in this group are not unlike those of the Adephaga in general appearance. They may be readily

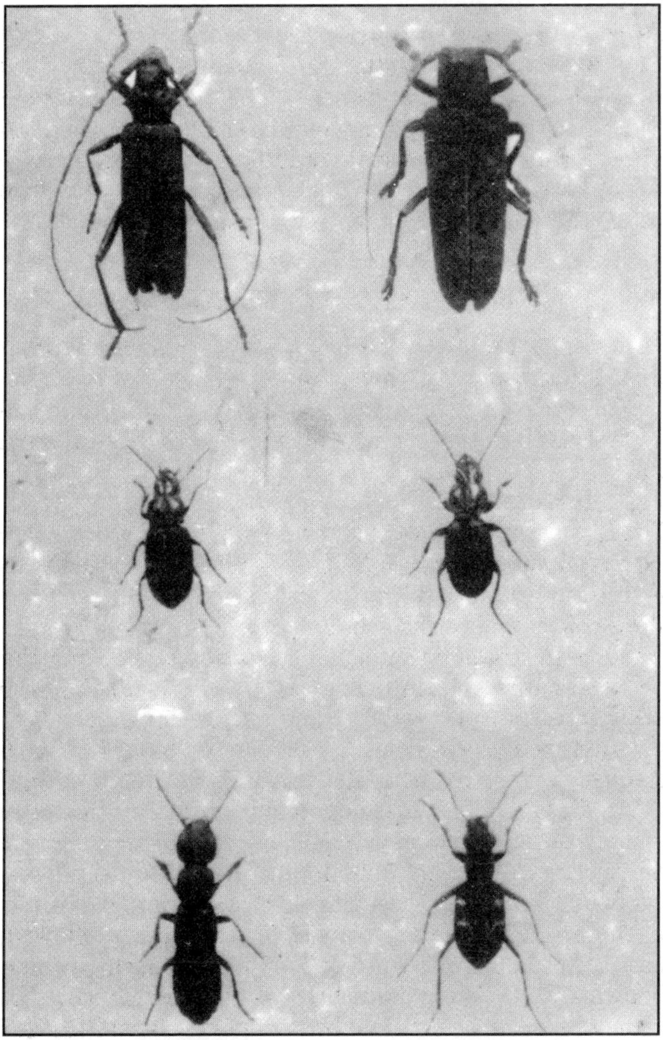

FIG. 69. -*Aromia moschata.* *Saperda carcharias.*
Harpalus ruficornis. *Pterostichus madidus.*
Ocypus olens. *Cicindela sy'vatica.*

distinguished, however, by the fact that they have only one claw on the legs whereas Adephagid larvæ have two. They are chiefly predaceous, but some feed on decaying matter.

FAM. SILPHIDÆ.

(Burying or Carrion Beetles).

The Beet Carrion Beetle (*Silpha opaca*) (Fig. 75), does damage to mangolds at times. The insects are taken to the field in farmyard manure. If this is ploughed under during autumn, the usual practice, it is unlikely that damage will occur, but if applied only a short time before sowing, then serious damage may result.

SUPER-FAMILY.—2. *Diversicornia*

In this group are placed a great number of families which differ widely in their characteristics. The majority are of no economic importance, at any rate in this country, but destructive species occur in the families mentioned below.

FAM. BYTURIDÆ.

The Raspberry and Loganberry Beetle (*Byturus tomentosus*).

Description and Life History.—The adult beetle is about one-sixth of an inch in length, and in old specimens the prevailing colour is dark brown (Fig. 70). Younger beetles are clothed with a golden or yellowish pubescence, but this in time wears off. During early May and throughout June beetles infest the flowers of raspberries, loganberries, wild and cultivated blackberries, and to some extent other fruit blossoms. They fly actively in sunshine but are sluggish on dull days. Eggs are deposited singly in the opening blossoms, where the larva feeds on the receptacle. More than one larva is sometimes found in each flower, presumably owing to more than one beetle having deposited an egg there. The larva when fully grown is yellowish in colour, and almost one-third of an inch in length. The head is brown and flattened, and the jaws well developed. Down the back are darker yellow-brown plates of chitin, and at the tail end a pair of cerci and an anal tube or pro-leg are visible. When fully fed, during July and August, the larvæ seek some place of shelter in which to pupate. Pupæ are found in the soil around canes, in cracks in the old canes, or in posts and stakes supporting the wires. There is only one generation per annum, for the pupæ which are formed during the summer remain in that stage until the following May, when the beetles again appear.

Plants attacked and damage done.—The beetle infests the flowers of raspberries, loganberries, and both wild and cultivated blackberries. It feeds in the adult stage on the flowers, causing many to fall, and partially destroying others. The larva attacks the developing fruit, boring into the receptacle and eventually also feeding on the little fruitlets. Migration of the larvæ may take place from one fruit to another, and if the attacked ones are not completely shrivelled and useless, they often become very soft and

FIG. 70. -*Phyllotreta nemorum* : *Byturus tomentosus* : *Meligethes aneus*.

unfit for transport and table use. So severe is the attack at times that the cultivation of loganberries has been threatened with extermination, and a large percentage of the crop is annually lost.

Control Measures.—It is important to remember that a plantation may become infested from wild blackberries in the vicinity, and where possible such bushes should be eradicated. The adult beetles may be collected on dull days on sacks or sheets put under bushes, which are tapped or shaken. Spraying with lead arsenate when the blossoms are about to open will poison many beetles when they commence feeding. A second spraying about two weeks later, when the flowers are more advanced has been found effective in destroying the larvæ, which devour the poison in boring their way into the receptacle. The use of arsenical sprays at this period may, however, poison bees which are visiting the flowers. It is important to cut out and burn all old canes so as to destroy pupæ over-wintering in them, and soil around the stocks should also be forked over. This disturbs and exposes pupæ to the influence of weather and the attacks of birds.

Poultry are very useful in destroying these beetles, as well as many others. Late hatched chicks can be run in a plantation with advantage, until they begin to do damage to the young leaves and blossom.

FAM. NITIDULIDÆ.

The Blossom Beetle (*Meligethes æneus*).

Occasionally some damage is done by this beetle and other species of *Meligethes* to cruciferous flowers. The beetles (Fig. 70) are about one-tenth of an inch in length, very dark blue in colour: they may be found in numbers in the blossoms of many Compositæ and Cruciferæ. They eat away the stamens and ovaries and may prevent the setting of seed. Eggs are laid in the blossoms and the larvæ also feed there. It is only occasionally that any great damage is done by them, and it is confined to crops from which seed is being raised.

FAM. COCCINELLIDÆ.

Ladybirds.

The majority of these insects must be reckoned as beneficial, since both the adults and the larvæ are predaceous.

Description and Life History.—Many species occur in this country, the two most common ones being *Coccinella septempunctata*, the **Seven-Spot Ladybird,** and *Adalia bipunctata*, the **Two-Spot Ladybird** (Fig. 71A). The latter often exhibits great variations in colour. Both these insects have a black head, very much sunken in the prothorax, which is of a similar colour. The elytra are red, and have seven or two black spots according to the species. The seven-spot ladybird is much the larger of the two. Both species are very convex in shape. Having hibernated in the adult stage, ladybirds appear in spring and early summer and lay their eggs in clusters in the vicinity of colonies of aphides, which constitute their chief food. The eggs are yellow, elongate, and similar in shape to a rifle bullet. They are laid with the long axis at right angles to the surface upon which they rest. In colour the larvæ are usually leaden black or grey, with yellow or reddish markings. They are commonly known among fruit growers as 'Niggers' (Fig. 71B). Like the adults they feed greedily upon aphides and scale insects (Coccidæ), and are of great service in keeping the numbers of such pests in check. As a rule, however, ladybirds do not appear until colonies of aphides are well established, and they cannot be relied upon to wipe out infestations of such pests.

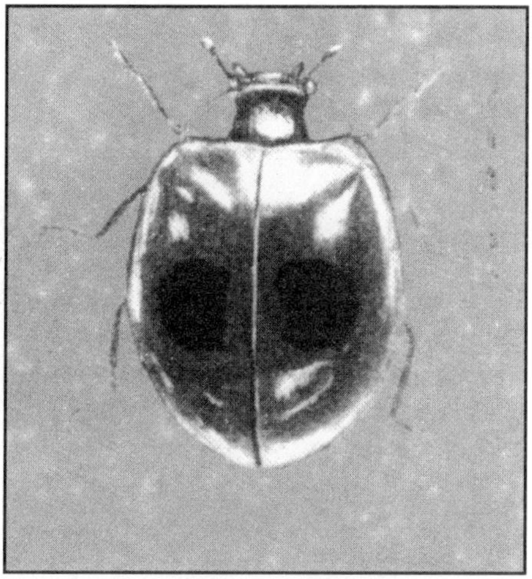

FIG. 71A. -Two-spot Ladybird Beetle (*Adalia bi-punctata*). Greatly enlarged.

FAM. DERMESTIDÆ.

The Larder or Bacon Beetle (*Dermestes lardarius*).

This insect is dark in colour, about half an inch in length, and covered with a grey pubescence (Fig. 75). Along with its larva, which is very hairy, it sometimes does damage to stored foods, especially fatty substances, but it is not to be considered as a pest of great importance.

FAM. ANOBIIDÆ.

Furniture Beetles.

Beetles of this family are responsible for the damage known as 'worm-eaten' furniture. The most common species in this country is *Anobium punctatum*, a small, narrow beetle about one-fifth of an inch long, and brown in colour (Fig. 72). The beetles appear in June and lay their eggs in cracks and crevices in the wood or in old burrows. Eggs are not laid on smooth or exposed surfaces. They are white in colour and lemon-shaped, and hatch in three or four weeks into a larva which bores its way into the wood. It continues to burrow its way along and does not come near the surface until it is about to pupate. Then it works outward until a very thin layer of wood remains between it and the surface. In such a position the pupa is formed. The pupal stage lasts two or three weeks, and the adult beetle on emergence gnaws its way through the thin layer of wood and comes out into the open air. The period occupied by the full life-cycle appears

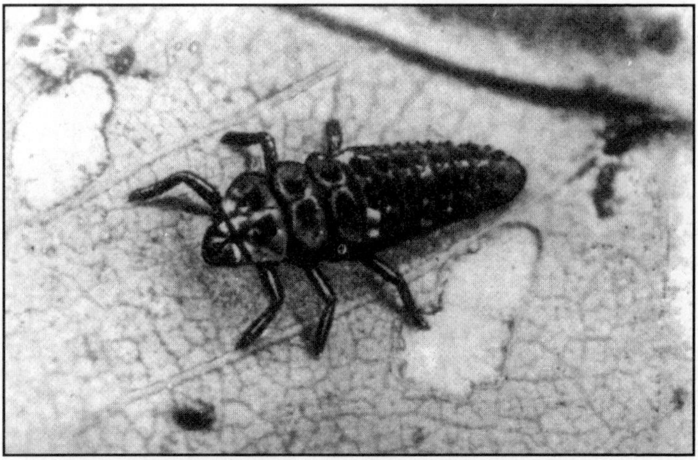

FIG. 71B. -Larva or "nigger" of Ladybird Beetle. Greatly enlarged.

to vary with the conditions. It may be completed in one year but often takes at least two.

Furniture in which holes are seen may have been made originally from wood which had been attacked. The only indication that larvæ are working in it is the appearance of fresh holes, and quantities of wood dust which come from them.

Hard woods are usually attacked in preference to coniferous woods, and the sapwood particularly seems to be the most attractive place for the beetles to commence their destruction.

Another beetle, commonly known as the **Death Watch Beetle** (*Xestobium rufovillosum*) (Fig. 72), causes similar damage to constructional timber, but does not appear normally to infest furniture. This beetle is almost a quarter of an inch long, and appears usually during April and May. In beams infested by this beetle a curious rapid tapping sound is sometimes heard. This is caused by the beetle striking its head on the floor of the burrow, and seems to be a sexual call since it only occurs just before the beetles emerge and pair.

Control Measures.—It is very difficult to deal with damage done by these beetles, but attacks may be largely avoided

FIG. 72. -*Anobium punctatum, Xetobium refovillosum, Anobium paniceum.*

by treating all unpolished surfaces, joints, etc., with paraffin or turpentine during the egg-laying season. Where it is possible to subject the wood to a dry heat, this method is very successful. The interior of the wood need only be raised to a temperature of about 55° C. Larger pieces of valuable furniture can be fumigated, but this treatment is dangerous unless carried out under careful supervision, and an air-tight chamber is necessary.

FAM. ELATERIDÆ.

Click Beetles.

This family contains some insects which are of great agricultural importance, the larvæ being the notorious 'wireworms.' In tropical countries some beetles of this family are the so-called 'fire-flies,' and they emit a striking phosphorescent glow at night. Most English species are dingy coloured, and the chief ones of importance are *Agriotes obscurus*, *A. sputator*, *A. lineatus*, and *Athous hæmorrhoidalis* (Fig. 73). *Agriotes obscurus* is probably the most common species, and its life history may be taken as typical.

Description and Life History.—The adult beetle is dull brown in colour and rather more than one-third of an inch in length. The antennæ are filiform and moderate in size. A large prothoracic plate and well-developed elytra cover the dorsal surface of the body. The legs are slender and rather small. On the under side of the prothorax (prosternum) is a backwardly projecting spine, and on the mesosternum an apparatus with which the tip of this spine engages. If the insect be placed on its back on a smooth surface, it engages the prothoracic spine and then frees it with a vigorous 'flick' of the head and prothorax. The result is that the insect is thrown up in the air and lands on its legs to the accompaniment of an audible click. This operation is responsible for the name of Click or Skip Jack beetles often given to this family. Adult beetles appear during the summer and fly readily on bright days. They may often be found on large umbelliferous flowers. Eggs are laid during the summer months in shady situations, chiefly under tufts of grass and just beneath the surface of the soil. Direct sunshine and dry conditions are fatal to the eggs and young larvæ, so that grassland, rough weedy banks and uncultivated pieces of ground are chosen for oviposition. On hatching, the tiny whitish larva immediately bores down several inches into the soil and appears to feed chiefly on decaying vegetable material. Under ordinary conditions the length of larval life seems to occupy from three to five years. Toward the end of its first year it appears to forsake feeding on humus and lives largely on the roots and shoots of plants near ground level. Its principal food-plants are grasses, but it is almost omnivorous and even predaceous at times. At this stage it takes on the appearance by which it is known to the farmer as a wireworm. It is a long, cylindrical grub, yellow or yellowish-brown in colour, and up to an inch in length. The brown, forwardly projecting head is provided with powerful jaws, and on each thoracic segment is a pair of legs. On the underside of the last abdominal segment is a projection known as the anal foot, and this segment tapers to a point at the extreme tail end. Wireworms belonging to the genus *Athous* are distinguished from this *Agriotes* type by reason of the fact that the hind end is not pointed but carries two

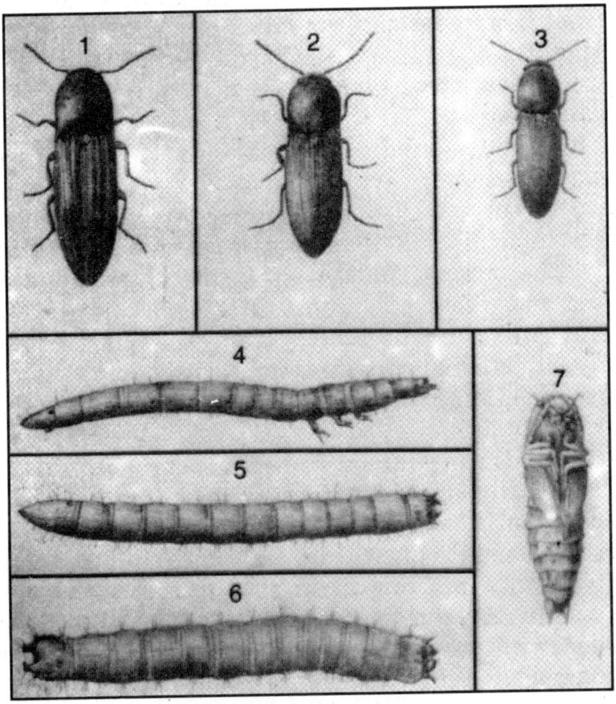

FIG. 73. -Click Beetles and Wireworms.

Fig. 1. *Agriotes lineatus*; 2. *A. obscurus*; 3. *A. sputator*; 4. *Agriotes* wireworm (side view); 5. *Agriotes* wireworm (upper surface); 6. *Athous* wireworm (upper surface); 7. Pupa of Click Beetle. All enlarged three times. Reproduced by permission of H.M. Stationery Office and the Ministry of Agriculture and Fisheries.

well-marked spines. When fully grown the wireworm constructs an earthen cell several inches below the surface of the soil and pupates. This takes place in the late summer. The pupa is white and soft and from it the adult emerges in about three weeks. Usually the beetle remains in the cell, or at any rate beneath the ground during the winter, emerging in the following spring.

Plants attacked and damage done.—It is easier to compile a list of plants which are seldom attacked than to enumerate those on which the wireworms will feed. In general, flax appears to be almost immune, beans and clover suffer little damage, and mustard and rape are not appreciably attacked. The chief loss to agricultural crops occurs on cereals, roots, and potatoes, whilst strawberries often suffer badly. Large numbers of wireworms are normally present in most grassland, but the regenerative power of grasses is so great that little actual damage is done. Particularly after grassland, including leys of two or more years' duration, is broken up, an attack of wireworms may be expected. It must be remembered that in such cases larvæ of various ages will be present, and an attack may therefore occur for the first few years of arable cultivation. Since few eggs are laid, however, in arable land, the intensity of the attack will gradually diminish. Eggs may be laid, however, in foul, weedy crops of any description, and trouble consequently arises.

Cereals.—During the latter half of April and throughout May, especially if the weather is unfavourable to growth, corn frequently becomes very yellow at the tips of the leaves and many plants die out entirely. Several consecutive plants in a row often fail, and the resulting crop is very patchy. Yellowing of the leaves should always be regarded with suspicion, since it is often an indication of root damage. Wireworms have a habit of working along a drill, boring into each plant in turn and eating out the base of the shoot. The leading shoots being thus killed a few weaker side tillers are produced, but a short and unevenly ripening crop is the inevitable result.

Potatoes.—When old or derelict grassland is broken up it is a usual practice in many districts to take potatoes as the first arable crop. If such land is badly infested with wireworms the 'sets' may suffer severe damage owing to wireworms boring into them, and also cutting through the sprouts. Weaker 'eyes' may be thus forced into growth, but cases have been seen of complete failure under these circumstances. When the crop is lifted the tubers are also found to be bored by wireworms, and a large percentage of such a crop is almost certain to be lost in clamping, by rots which are set up.

Roots.—Mangolds and swedes are sometimes severely damaged in the seedling stage when wireworms are numerous. Damage is caused by the larvæ eating at the root of the seedling plant, and many birds, including rooks and starlings, cause further loss by pulling up plants in their search for the wireworms.

Glasshouse Crops.—Under conditions of intensive cultivation, wireworms are often introduced into glasshouses in turf. An unusual kind of attack is sometimes seen on tomatoes, the larva boring into the stem below ground and working up inside it to a height of several inches.

Control Measures.—At the present time, when conditions are such that very little grassland is being brought under the plough again, complaints of wireworm damage are not so frequent as formerly. During the Great War a great deal of old turf was ploughed up for wheat growing, and very serious damage was done. It is not uncommon for a field to have as many as 250,000 wireworms per acre; under good conditions of cultivation it is possible that crops will not be greatly damaged unless the number of wireworms is somewhere in this neighbourhood. Where possible it is better to leave a ley down for one year only, since under such conditions wireworms seldom become numerous enough to cause much damage. If, however, a longer term ley is ploughed up, or old grassland brought into cultivation again, it is essential that the turf be well buried. Cereals in particular suffer from wireworm when they are checked from any other cause, and one of the difficulties with ploughed-up turf is to consolidate the ground sufficiently. Hollows between the furrows result in a loose seed-bed; the seedlings get 'hung' and fail to root and grow well. Plants in this condition suffer most, and any means by which a good firm seed-bed can be obtained tend to encourage the plant and restrict the movement of the wireworms. For this reason, especially on spring-sown cereals, heavy rolling has been found most beneficial on land which will not suffer with such treatment. In order to facilitate consolidation of the land, old grassland should be broken up as early as possible in the summer. If a hay crop is taken, then the land should be ploughed immediately the hay is carried : or if grazed, the field should be grazed 'bare' and then broken up. Bastard fallowing of this kind does a great deal towards ensuring a successful crop. It is sometimes stated that if a crop of mustard be grown and ploughed in, wireworms can be almost eradicated. Too much weight should not be attached to such statements, since a great many other conditions may have influenced the result. Similarly it has been averred that rape cake attracts wireworms and that they will feed on this substance instead of attacking growing plants. Recent work, however, has shown that germinating cereals attract wireworms more strongly than any such baits. Many brands of soil insecticides are said to kill wireworms. Their efficiency under field conditions has not been conclusively proved, and there is little doubt that the most remunerative control methods are those of a cultural nature. Great help may be given to the plant by suitable nitrogenous top dressings in the spring, and the crop thus pushed beyond the danger stage.

In the case of oats those varieties which possess a good tillering power should be selected for sowing on infested land.

Potatoes are usually a fairly safe crop unless very large numbers of wireworms are present, but little can be done to check the damage. Where roots are attacked starlings and rooks often congregate in large numbers, and if surface cultivation is carried out regularly and at short intervals, they will account for many of the wireworms. Lapwings and pheasants are also beneficial in this way. Under conditions of intensive cultivation, such as market gardening, it may be considered worth while to take active measures to kill the wireworms. It has been found that they are strongly attracted to germinating cereals, and when congregated together in this fashion may be more easily destroyed by a soil insecticide. Recent experimental work has demonstrated the possibility of effecting a control by sowing wheat in rows several feet apart, and following some days later with calcium cyanide drilled down the rows a few inches beneath the soil. Hydrocyanic acid gas is evolved, which is rapidly fatal to the larvæ. On a small scale, in glasshouses, wireworms are frequently trapped by burying split carrots just beneath the soil at intervals. These baits are periodically examined and the wireworms which have congregated there are destroyed.

FAM. CRYPTOPHAGIDÆ.

The Pigmy Mangold Beetle (*Atomaria linearis*).

This tiny beetle is less than 2 mm. long, parallel-sided, and of a reddish-brown colour (Fig. 74A). It is found in numbers in haystacks and among heaps of refuse. Details of its life history are not known, but the larvæ are sometimes found around the roots of seedling mangolds and sugar beet, upon which they feed. The tap root, below ground-level, turns black and shrivels into a thin stringy fibre, while the crown of the plant develops almost normally (Fig. 74B).

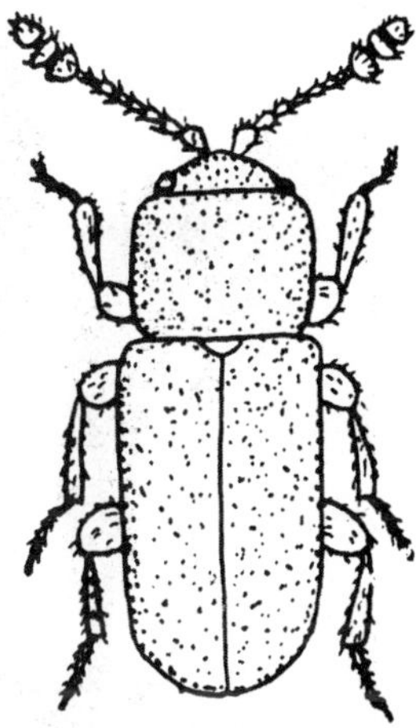

FIG. 74A.—The Pigmy Mangold Beetle (*Atomaria linearis*). Greatly enlarged.

Should weather conditions be unfavourable to the growth of the crop, many seedlings may be killed out as a result of the attack. Crops on the lighter soils appear to suffer most. At present no means have been found of controlling this insect.

Damage of a somewhat similar nature is caused by springtails (Collembola) (p. 47). *Atomaria*, however, always attacks the root below ground-level, whereas springtails feed at, or just above, the surface of the soil.

SUPER-FAMILY—3. *Heteromera*

In this super-family there are no insects of economic importance in this country, with the exception of some in the family *Tenebrionidæ*. The larva of the common species, *Tenebrio molitor* and *T. obscurus*, are sometimes found in food and stored grain products, and are known as Meal-worms. They are very similar in appearance to wireworms.

SUPER-FAMILY—4. *Phytophaga*

FAM. BRUCHIDÆ.

Members of this family are frequently found in legumi-

FIG. 74B. -Mangolds damaged by Pigmy Mangold Beetle.

nous seeds, particularly peas and beans, on which the larvæ feed. They are often called 'pea and bean weevils,' but this name is quite erroneous (*see* Rhyncophora).

Bruchus pisi and *Bruchus rufimanus*.

Description and Life History.—In general appearance these two beetles closely resemble one another. They are about one-sixth of an inch in length, black in colour, and covered with a brown pubescence. The four basal joints of the antennæ are reddish and the tibiæ and tarsi on the first two pair of legs are of a similar colour. In *B. pisi* the femora of the front legs are black, whereas in *B. rufimanus* the same joints are red. The elytra are rather short, exposing the terminal segments of the abdomen (Fig. 75).

Eggs are laid on the very young pods, those of *B. pisi* on peas and of *B. rufimanus* on beans. On hatching, the larva bores into the developing seed, where it lives until fully grown. It has three pairs of thoracic legs when very young, but these are lost eventually, and the fully grown larva is a legless, yellowish-white and much wrinkled grub. The brown head is provided with powerful mandibles. Pupation takes place inside the seed just beneath the seed coat, through which the adult beetle gnaws a circular hole when it emerges. Emergence may not take place until the spring, after the seed has been sown. The adults are capable of flying considerable distances.

FIG. 75. -(Top) *Bruchus pisi* ; *B. rufimanus* ; (centre) *Leptinotarsa decemlineata* ; (below) *Dermestes lardarius* ; *Silpha opaca*.

Plants attacked and damage done.—*Bruchus pisi* seems to confine itself to peas, while *B. rufimanus* is found chiefly on beans. The beetles themselves are said to cause some damage by feeding on the leaves of the growing plants, but the main damage is caused by the larvæ feeding in the seed. Attacked peas and beans have a lower market value than sound ones, and their germination may be seriously impaired. It is not easy to detect attacked seeds until the holes made by the adults are seen. In the case of beans a round, light-coloured area is sometimes visible, and if this be opened the larva or pupa will be found beneath. On heavy land, where beans are grown, the Wild Oat (*Avena sativa*) is often a troublesome weed, and it is recorded that the grains of this weed become fixed in the holes in the beans, so that the two are sown together.

Control Measures.—Infested seed should not be sown. Its germination is often poor, and beetles emerging from it will naturally attack the new pods. If the seed be thrown into water healthy peas will generally sink at once, but attacked ones often sink also in a very short time, so that this method of separation is not to be relied upon. Fumigation by carbon disulphide is the most effective way of dealing with the pest.

FAM. CHRYSOMELIDÆ.—This large family includes a number of very injurious insects. Most of them are small or only moderate in size, and the majority are characterized by bright metallic colouring, particularly on the elytra. Many of the species pass over the winter in the adult stage, the beetles hibernating in sheltered places.

The Asparagus Beetle (*Crioceris asparagi*).

Description and Life History.—The beetle is about three-quarters of an inch long, slender in form, and brightly coloured. The head is black and the thorax red, with a few black lines. The elytra are variously marked with yellow and black, and the appearance of these markings is that of a cross (Fig. 76). Very elongate, oval eggs about one-sixteenth of an inch in length, are laid in rows on the 'heads' of foliage of the plant. They are greenish-brown in colour and covered with an objectionable gummy secretion. The larvæ are yellow or dull green in colour, and almost half an inch long when fully fed. The head is black, three pairs of thoracic legs are present, and there is an anal pseudopod.

Eggs are laid during June upon the heads of asparagus, and the larvæ feed there, causing brown, discoloured patches, and soiling the heads with a gummy secretion. Larvæ become fully fed in about two weeks. Pupation takes place in the soil or beneath surface rubbish, and lasts from two to three weeks. There may be three broods per annum under favourable conditions, the adults of the last one hibernating in hollow stems, under the bark of trees, and in such-like places.

Natural Enemies.—Ladybirds and their larvæ and also the larvæ of the Lacewing Fly feed upon the larvæ of this beetle and serve to check its increase.

Control Measures.—Young beds of asparagus are most liable to infestation, and every effort should be made to prevent the pest becoming established. Hand-picking is

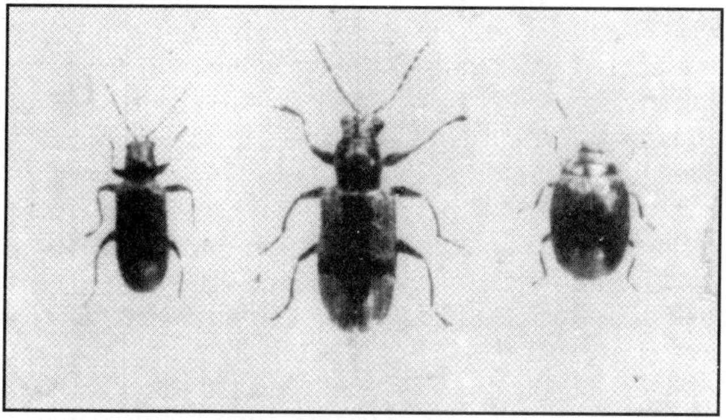

FIG. 76. -*Lema melanopa* ; *Crioceris asparagi* ; *Phaedon corhleariae*.

the most efficacious proceeding on a small scale, or finely powdered lime may be dusted on.

A few heads may be left to serve as a trap, since the beetles lay their eggs on these in preference to other growth. Such heads should be cut and destroyed regularly.

Paraffin and soft-soap emulsion is effective as a spray, and is useful if frequent applications are made.

Lema melanopa (Fig. 76).

The slug-like larvæ of this small beetle sometimes do considerable harm by stripping the epidermis from the blades of oats. Long brown stripes appear on the leaves, and attacked plants may be seriously stunted. The larvæ are about a quarter of an inch long, dark brown in colour, and slimy to the touch. The insect appears in serious numbers at times, but it is generally distributed and has not been fully investigated.

The Colorado Beetle (*Leptinotarsa decemlineata*).

This insect is a pest of potatoes in America, and lives also on allied plants. In general appearance it is somewhat like a large Ladybird, with yellow elytra, each of which is marked with five black lines (Fig. 75). It has at times been introduced into this country, but drastic measures have been taken to exterminate it at once. All precautions are now taken at ports to prevent its reintroduction, and any appearance of the insect must be immediately notified.

The Mustard Beetle (*Phædon cochleariæ*).

Particularly in the Fen area of England, where mustard is grown for seed, this beetle is sometimes the cause of serious loss. A closely allied species, *P. armoraciæ*, is often present also, and another, *P. tumidulus*, has been recorded as attacking various field crops, particularly carrots.

Description and Life History.—P. cochleariæ is a brilliant metallic-blue beetle, very convex in shape, and about one-eighth of an inch long (Fig. 76). Hibernation takes place in straw and dry refuse. The beetles emerge in spring, and often fly long distances to their food plants. Mustard crops in rough leaf are attacked by this generation, but the attack is not usually serious, since the crop grows very rapidly. Eggs are laid beneath the leaves during May and June. The larvæ are fully fed in two or three weeks, when they fall to the ground and pupate. A second generation of beetles appears, eggs are laid again on the leaves, and a further emergence of beetles takes place about August. The plant is then setting seed, and there may be huge numbers of beetles on the crop. These go into hibernation with the approach of winter.

Plants attacked and damage done.—Mustard is the chief crop attacked, but once this has been cut the beetles may migrate to neighbouring cruciferous crops. Beds of water-cress are sometimes badly damaged unless they can be flooded. Little appreciable damage is done to mustard when it is in full growth. The critical period is that just prior to the ripening of the seed, when huge numbers of the beetles are often found stripping the bark and causing great destruction, assisted by some larvæ still feeding on the leaves. Almost total loss of the seed crop may result, unless the weather is favourable to quick ripening. A few days' difference in the time of ripening of the crop may mean success or failure.

Control Measures.—It is almost impossible to deal with an attack of the beetle when the crop is ripening, and efforts should be directed towards checking the early generations. Spraying with a stomach poison can be carried out when the plants are small, and beetles can also be destroyed by machines designed for 'sweeping.' Any measures tending to reduce numbers during the early summer will help to obviate the later and more serious attack. To this end hedge bottoms and the banks of dykes should be kept as clear as possible from accumulated rubbish, and the straw from the mustard crop, which has little commercial value, is best burned, since it provides hibernation quarters.

The Turnip Flea Beetle (*Phyllotreta nemorum*).

Along with this species several others are often found attacking cruciferous crops; these are *Phyllotreta undulata*, *Phyllotreta cruciferæ* and *Plectroscelis concinna*.

Description and Life History.—Flea Beetles are small insects, not more than one-tenth of an inch long (Fig. 70). *Phyllotreta nemorum* and *P. undulata* are black in colour, with a yellow longitudinal stripe on each elytron. *P. cruciferæ* is uniformly metallic blue and *Plectroscelis concinna* brassy black.

The characteristic of all these beetles is their power of jumping. The femora on the metathoracic legs are very strongly developed, and on being disturbed the insects jump several inches. At times, when danger threatens, they also sham death, and lie inert and seemingly lifeless on the ground. During fine weather they fly readily and may migrate for considerable distances. The winter is passed by adult beetles in hibernation. Great numbers are found at times in stack bottoms, heaps of refuse, and dirty hedgerows and banks. In spring the beetles leave these quarters and migrate to cruciferous crops, on which they feed, making characteristic holes in the leaves, particularly the cotyledons. The exact life history is still in doubt in the case of some species. Eggs are generally laid on the leaves, but may also be found in the soil near the plants. Larvæ are found both mining in the leaves and feeding, just beneath the soil, on the roots. Pupation always takes place in the soil. The total time occupied by the egg, larval and pupal stages, varies a good deal. It may be completed in about six weeks or may take as long as ten weeks. The adult beetles which emerge during the summer live through the winter, and well into the following summer.

Plants attacked and damage done.—Almost all cruciferous plants, both wild and cultivated, form the food plants of these Flea Beetles. They are found very commonly on charlock. Not only field crops of swedes and kale, but garden crops, such as cabbage, broccoli, etc., are often badly attacked. The damage is chiefly done to the cotyledon leaves, but the plants may suffer severely even when in rough leaf, during adverse weather conditions. Hot, dry spells of weather result in the plants making very little growth, and at such times the beetles appear to be particularly numerous. Two or three successive sowings may be cleared unless measures are taken to check the attack. Both the adult beetles and their larvæ cause damage, but the adults in particular are the worst enemies. The epidermis of the leaves is stripped off over small areas and often holes bitten completely through. *Plectroscelis concinna* has been recorded as attacking a variety of other plants beside crucifers. Mangolds, hops and barley are sometimes damaged by it.

Control Measures.—The most dangerous period is that during which the crop is in the seedling or two-leaf stage. Once the plant begins to form its 'rough' leaves, the chances of injury are not so great. It is therefore obvious that any means which will assist the rapid growth of the plant in its early stages will be of great benefit. Correct cultivation is absolutely essential. One vital factor affecting the growth of the seedling swede or kale plant is water, and in order to ensure this the seed-bed needs to be both fine in texture and firm, after having been cultivated deeply and thoroughly. Farmyard manure is of course almost essential on most soils. A fine tilth has another advantage also, in that it affords less hiding for the beetles. Damage is invariably greater on a rough seed-bed, and even in rough patches in the same field, than on the finer tilth. In spite of all cultural operations, however, cold rainless periods often check the growth of the plant during the critical stage, and other measures must be employed to lessen the risk of injury. Various forms of trapping or sweeping devices have been employed with success, and a machine is now on the market which is designed for large-scale work. The essential feature of all such machines is a tarred or greased board, carried just above the plants, so that the beetles adhere to it when they jump. Such machines must be pushed and not drawn (Fig. 28). If men or horses go in front and pull it, they disturb the beetles, which jump away. The first thing which disturbs the beetles must be the greased board or tray.

Flea Beetles are intolerant of dust, and it has been found that continual scuffling of the crop, or even driving sheep rapidly through it, will often stop an attack. Basic slag, or other fine powder, dusted on during the early morning when the plants are wet with dew is frequently effective.

As a preventive measure, it is said that dressing the seed with paraffin or turpentine will ward off the attack. About 7 lbs. of seed can be dressed with a dessert-spoonful of the liquid just before sowing. The efficacy of such a measure, however, is uncertain, and its value has not been definitely established.

On small garden or market garden areas, spraying with lead arsenate is quite effective.

Weeds, such as charlock, are food plants of these beetles, and they will feed on them as well as on crop plants. All such weeds should therefore be kept down.

Burning of useless stack bottoms and accumulations of dry grass and other vegetation is helpful in destroying likely hibernation places.

FAM. CERAMBYCIDÆ.

Longicorn Beetles.

This family contains a number of beetles whose larvæ are wood-borers. The adults in many cases are large, brightly coloured insects with conspicuous long, simple antennæ (Fig. 69). *Saperda carcharias* and *Aromia moschata* are two common species, whose larvæ bore into trees, particularly willows.

The larvæ possess very minute thoracic legs, and the abdominal segments carry swellings both on the dorsal and ventral surfaces to assist locomotion in the burrows. The head is small and furnished with mandibulate mouth-parts.

Injury is caused to growing trees, but it is not as a rule severe, and no active measures have been taken to prevent it.

SUPER-FAMILY—5. *Rhyncophora*

The majority of these insects are easily recognizable by reason of their somewhat pear-shaped bodies and the possession of a marked rostrum or proboscis. This rostrum is in no way associated with the mouth parts; it is a prolongation of the front of the head, and the mouth parts are situated at the end of it. It is not equally well developed in all families, and it may take the form of a slightly elongated head in some, while in others it is as long as the remainder of the body. The antennæ are of the geniculate-clavate type, and are usually prominent organs. The larvæ are almost always legless, with a well-developed head and a much-wrinkled body.

FAM. CURCULIONIDÆ.

Weevils.

This is by far the largest family in the Rhyncophora, in fact it is said to be the largest natural family in the animal kingdom. Insects belonging to it are commonly known as Weevils. A number of them are injurious to cultivated plants and trees, their importance being horticultural rather than agricultural.

The Black or Vine Weevil (*Otiorrhynchus sulcatus*).

Along with *O. picipes*, the **Raspberry or Clay-coloured Weevil,** and *O. tenebricosus*, the **Red-legged or Plum Weevil,** the above species is sometimes responsible for much injury to plants.

Description and Life History.—In no species of *Otior-*

rhynchus is the rostrum well developed, but the forepart of the head is rather flattened and elongate (Fig. 77). These weevils are unable to fly, the elytra being secured to the abdomen.

O. sulcatus and *O. tenebricosus* are almost half an inch in length, and *O. picipes* about one-third of an inch long. The eggs of all species appear to be laid in the soil during the summer, and the larvæ feed upon plant roots, including strawberries and a variety of perennial and glasshouse plants. Adults emerge in the spring and feed upon buds, leaves and young shoots, of which the bark is 'ringed.' Vines, peaches, nectarines, raspberries and plums suffer most, and damage is often caused to apple grafts.

FIG. 77. -(Top) *Otiorrhynchus tenebricosus*; *O. Sulcatus*; (bottom) *Liophlaeus nubilis*; *O. Picipes*.

All are nocturnal feeders, and hide chiefly in the soil by day.

Control Measures.—When disturbed these beetles drop to the ground and feign death. They are very difficult to detect until they move again, but if tarred or greased sheets are spread beneath infested trees before they are tapped or shaken, numbers of the weevils may be caught. Such treatment must be carried out at night and should be repeated several times.

Pea and Bean Weevils (*Sitones lineatus* and *Sitones crinitus*).

Description and Life History.—Both these species, and sometimes others, feed on leguminous plants, particularly peas and beans. They are small, brownish-coloured weevils nearly a quarter of an inch in length, and have no marked rostrum (Fig. 78).

Eggs are laid in spring just beneath the soil around leguminous plants. The larvæ are creamy white and about a quarter of an inch long when fully grown. They have a well-developed, reddish-brown head, a powerful pair of jaws, and are legless. Their food consists of the roots, along which they work, and the root nodules are particularly attacked. Pupation takes place in a cell about two inches below the surface of the soil. Adult beetles appear in the summer and many of these hibernate until the following year. Some, however, appear to lay eggs during the same summer, and the larvæ live in the soil, particularly at the roots of clover, throughout the winter. Beetles which have hibernated emerge in the spring and at once commence to feed on peas and beans, cutting characteristic semicircular notches in the margins of the leaves (Fig. 79). When disturbed these weevils fall to the ground and feign death. During dry weather crops are severely damaged and may be complete failures.

FIG. 78. -(**Top**) *Sitones lineatus*; *S. Crinitus*; (centre) *Anthonomus pomorum*; (bottom) *Phyllobius maculicornis*; *P. Oblongus*.

Plants attached and damage done.—Peas and beans suffer most, especially on a rough tilth and when conditions are not in favour of their rapid growth. Red clover is often badly damaged at the roots, and the result of larval attacks is often a poor second cut.

Control Measures.—The importance of a good tilth and suitable manuring cannot be over-estimated where crops of peas and beans are grown in gardens or market gardens. Finely powdered substances such as soot and lime, applied when the plants are wet with dew, deter the beetles, and a stomach poison spray, such as lead arsenate, is also effective. It is difficult, however, to cover the leaves of peas satisfactorily since the spray fluid will not adhere to them readily. The burning of accumulations of rubbish is to be recommended during the winter, as many hibernating beetles will then be destroyed.

FIG. 79. -Pea Plant damaged by *Sitones*.

The Apple Blossom Weevil (*Anthonomus pomorum*).

Description and Life History.—The adult beetle is about a quarter of an inch in length and has a well-developed rostrum. Its general colour is brownish, but there is a very broad, lighter V-shaped mark on the elytra (Fig. 78). In spring the weevils crawl or fly on to apple-trees and feed to a small extent. When the blossom buds are developing and before they have opened, the female weevil bores a hole in them and deposits an egg within the bud, sealing up the hole with saliva. Only one egg is laid in each bud, but each beetle may lay upwards of fifty eggs. A small white legless larva hatches and immediately begins feeding inside the bud, which is prevented from opening. If the weather is very favourable and the blossom opens rapidly, many of the very young larvæ perish before they can do much damage. Attacked buds, however, fail to open, and the petals turn brown and die. Such unopened and dead buds are called 'capped blossom' and are very characteristic of the weevil attack (Fig. 80). The larva when fully developed pupates inside the bud, from which the weevil emerges by

FIG. 80.—" Capped " apple blossom due to attack by Blossom Weevil.

boring a circular hole. The adults pass the winter in the shelter of rough bark and under dead leaves, etc.

Plants attacked and damage done.—Apples only are attacked, and damage has been reported from all over the British Isles. It is most severe in the South Midland and Southern counties. When there is a plentiful blossom and the fruit sets well, the buds killed by the weevil only serve to produce a useful thinning of the crop. In adverse seasons, however, the attack is serious and results in a depleted crop.

Control Measures.—So severe have the attacks of this weevil been on the Continent that hand-picking of the 'capped blossom' has been resorted to in places. Such blossoms are easily seen and their destruction lessens the numbers of adult weevils. Such a procedure needs co-operative work over any given district, since the adults are capable of active flight.

Loose bark and refuse heaps form the chief hibernating places for the adults. Artificial hibernation quarters in the form of sack or hay bands round the trees attract many beetles if the bands are in position before winter. These bands should be removed and burned before spring.

A hymenopterous parasite, *Pimpla pomorum*, attacks the larva of the Apple Blossom Weevil. The parasite lays an egg in the capped blossom, and the larva, on hatching, feeds upon and kills the weevil larva. If 'capped blossoms' are collected and kept, a number of the adult parasites will hatch out, and these may be liberated again, while the adult weevils should be destroyed.

Leaf Eating Weevils (*Phyllobius pyri* and *P. oblongus*).

Many species of *Phyllobius* occur on various trees and devour the foliage. They are all medium or small insects with no marked rostrum and are often brightly coloured (Fig. 78). At times the foliage of fruit trees, particularly plums, suffers severe damage. Lead arsenate spray will destroy the beetles.

Pear-shaped Weevils (*Apion apricans* and *A. pomonæ*).

These tiny blue or purplish weevils sometimes attack clovers, lucerne and other Leguminosæ. They have a curved rostrum and are sometimes known as 'Pear-shaped Weevils' (Fig. 81). They are about one-eighth of an inch in length and often occur in great numbers.

The Granary Weevil (*Calandra granaria*).

This is a small, dark-coloured weevil, less than a quarter of an inch in length (Fig. 83). A similar species, *Calandra oryzæ*, is sometimes found in company with it. Both these weevils are now widely distributed, and their larvæ destroy quantities of cereals, rice, etc., in storage. Severe infestations in mills and similar buildings have to be dealt with by fumigation.

The Root Gall Weevil (*Ceuthorrhynchus pleurostigma*).

The roots of many Brassicæ, especially cabbages, cauliflowers and swedes, are sometimes affected by round galls about the size of a pea, or rather larger. If such galls be opened a small larva is found within. It is of the typical weevil type, with a brown head and without legs. Such swellings on the roots are often mistaken for an attack of the Club Root or 'Finger and Toe' fungus, but the presence of the larva at once determines the cause (Fig. 82). The adult weevil is a small dark insect (Fig. 81), and the eggs are laid on the roots of the plants. The larval feeding causes the root tissue to grow

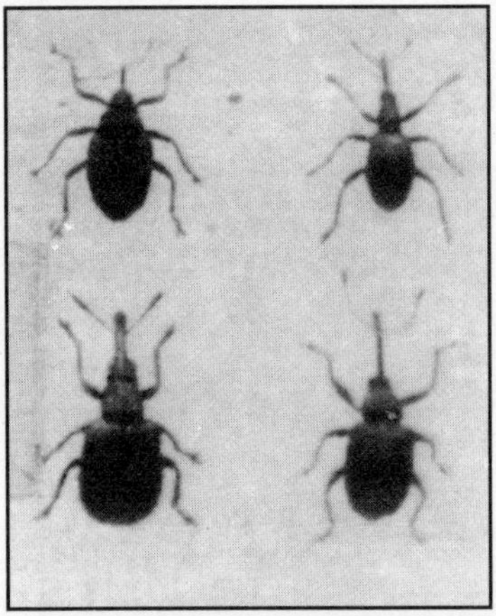

FIG. 81.—*Apion apricans; Ceuthorrhynchus pleurostigma* (above); *Anthonomus rubi; Rhynchites cæruleus* (below).

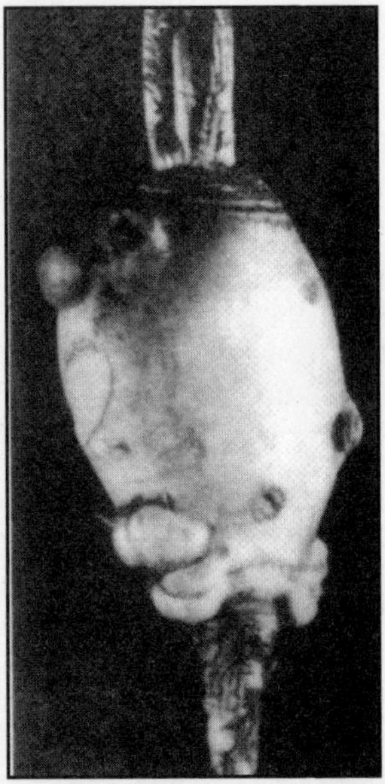

FIG. 82.—Galls on Turnip caused by Turnip Gall Weevil.

out abnormally into galls which envelop the larva. Although this insect is very prevalent in some localities it does little damage, at any rate to strong growing varieties under good cultural conditions.

FAM. SCOLYTIDÆ.

Bark Beetles.

This family, which is very closely allied to the weevils, contains a number of beetles responsible for damage to the wood of trees. They must not be confused with the previously mentioned Furniture Beetles (Anobiidæ).

Description and Life History.—A number of species occur in England and all the adults are of small size and somewhat cylindrical shape. There is no rostrum as in the true weevils. Three of the commonest species are *Scolytus rugulosus*, *S. pruni* and *Xyleborus dispar* (Fig. 84). In colour they are dark, sometimes shining black, with a reddish tinge on the elytra. The insects bore holes into the bark of the tree, in some species extending only as far as the wood and in others penetrating deeply into it. In this gallery the female lays a number of eggs at intervals and the larvæ, on hatching, bore other galleries which radiate out from the parent one. In the blind enlarged ends of the galleries the larvæ pupate, and the adult beetles bore their way to the exterior. In some cases there may be two broods in a year.

FIG. 83.—*Calandra granaria. C. oryzæ.*

Plants attacked and damage done.—Elm, oak, ash, pine and fruit trees are commonly attacked. Damage is done particularly to felled timber or broken boughs, though living but unhealthy trees may be attacked. In some instances the larvæ have been found to be destroyed by sap exuded from the tree flooding the galleries.

Control Measures.—All fallen and dead wood should either be barked or cleared away as soon as possible, especially the larger boughs, in which these beetles chiefly breed.

SUPER-FAMILY—6. *Lamellicornia*

The beetles in this family have their antennæ in the form of a lamellate club (see Fig. 68), which makes them easily recognizable. The Lamellicornia includes the largest beetles in the world, which are tropical in distribution.

FAM. SCARABÆIDÆ.

Chafers and Dung Beetles.

The Cockchafer (*Melolontha vulgaris*).

The adult beetle is about an inch in length. The head and prothorax are black, and the elytra brown and covered with hairs. On each elytron there are five longitudinal ridges. The terminal segment of the abdomen is elongate, curved ventrally, and not covered by the elytra. The male has seven laminæ in the antennal club and the female six. When fully grown the larvæ are about one and a half inches long, creamy white in colour, and with the abdomen curved ventrally in characteristic fashion and somewhat swollen at the tip (Fig. 85). The head is chestnut brown. There are three pairs of legs on the thoracic segments. The legs are also brownish in colour and well developed, the metathoracic pair being the longest. The larva constructs an earthen cell several inches below the surface of the soil, in which it pupates. The pupa is creamy white in colour. Eggs are laid during the summer chiefly in grassland, the female burrowing into the earth for the purpose. About seventy eggs are deposited in several batches. These hatch in five or six weeks, and the larvæ feed on the roots of grasses and seedling trees. The exact time occupied in the larval stage is uncertain, but is probably four years. At the end of its fourth summer the larva pupates but the beetle does not emerge until the following year. Adults are, therefore, only prevalent every fifth year, and these are known as 'flight years.' A few stray beetles may, however, be seen in most years. The adults fly readily in the dusk, when large numbers congregate around tree-tops, particularly oaks. The beetles feed to some extent on the leaves and remain concealed in the trees by day. A similar but rather smaller beetle is the **Summer Chafer** (*Rhizotrogus solstitialis*), which appears to complete its life-cycle in one year. It may be distinguished from the Cockchafer by having four ridges on each elytron and only three laminæ to the antennal club in both male and female.

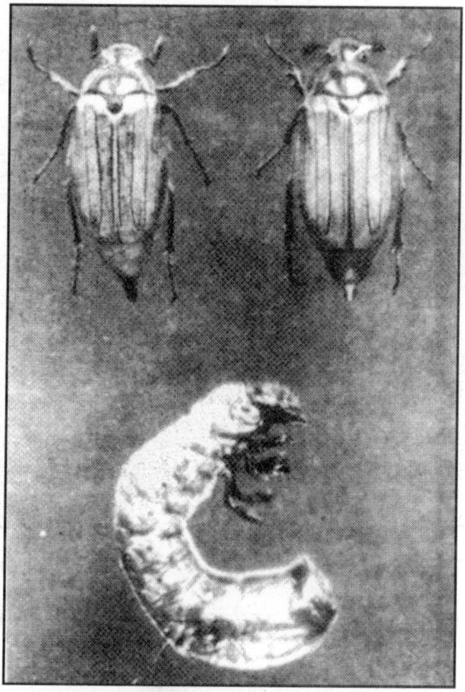

FIG. 85.—Cockchafer (*Melolontha vulgaris*)—male, female, and larva.

A still smaller insect is the **Garden Chafer** (*Phyllopertha horticola*), which is about half an inch in length. The **Rose Chafer** (*Cetonia aurata*) has elytra of a brilliant golden green colour, and it is about the size of the Summer Chafer (Fig. 86).

Plants attacked and damage done.—The natural food plants of chafer larvæ are grasses, but many other plants are attacked, especially cereals. Seedling trees also are often severely damaged. The attack takes place at the roots and the larvæ never come above ground. Cereals are very rapidly destroyed, and grassland shows large, brown, dying patches where the larvæ are most numerous. Occasionally root crops are attacked to some extent if sown on fields in which the larvæ are present, and cases of damage to sugar beet have been observed. Certain fields appear to be more liable to such attacks than others, and the proximity to belts or plantations containing a number of oak-trees seems to predispose a field to infestation. As with the click beetle, so the eggs of the cockchafer are laid in grassland and the larval damage is often attributed to wireworms.

Natural Enemies.—Rooks and other birds feed readily on these larvæ and congregate in flocks when an infested field is being cultivated. Gulls and plovers, where they occur, are also very beneficial birds. Moles devour the

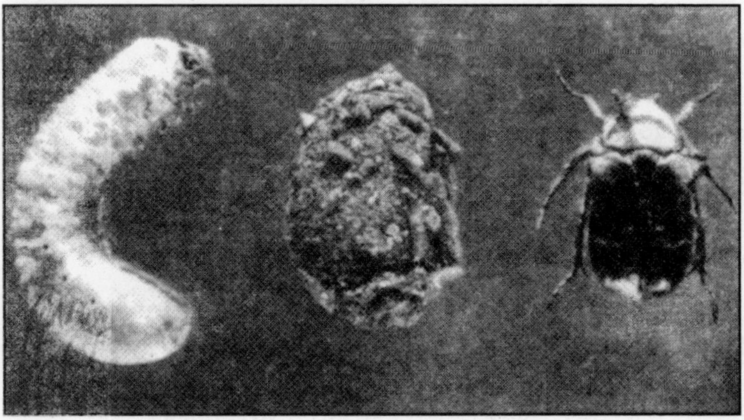

FIG. 86.—Rose Chafer (*Cetonia aurata*)—larva, cocoon, and imago.

grubs readily, and bats, owls and nightjars feed upon the adult beetles.

Control Measures.—Grassland: naphthalene and insecticides containing that substance are effective against chafer larvæ. Dressings of two hundredweight or more per acre should be applied. The toxicity of naphthalene is more pronounced when the application is followed by rain which washes it well into the soil. Under dry conditions, results are liable to be disappointing. Very heavy rolling also destroys the larvæ and a steam-roller has been used with great advantage. Drastic as this method might appear to be, it has been found to be cheaper than other operations and considerably more effective.

Arable Land.—In growing crops little can be done. When

land is being cultivated the larvæ are readily seen, and precautions should be taken to get rid of them before sowing. It is both cheaper and more effective to carry out extra cultivations, which expose the larvæ to the attacks of birds, than to resort to the use of insecticides.

On a small scale the grubs are readily trapped by pieces of turf buried a few inches beneath the surface. Such traps should be periodically examined and the grubs destroyed.

CHAPTER XV

HYMENOPTERA—ANTS, BEES, ICHNEUMON FLIES, ETC.

THE best-known insects in this order are the ants, bees and wasps, which have developed social habits to a very high degree. They live in large colonies, and have male, female and worker forms or castes. Occasionally other specializations are met with. The instincts of these insects are very highly developed, and they have consequently attracted much notice. By no means all the Hymenoptera fall into these families, however. A vast number of insects are known which are parasites, chiefly upon other insects, and these are sometimes grouped together as the *Parasitica*. On the whole their activities are beneficial from man's point of view. There is also a group known as Sawflies, whose larvæ somewhat resemble those of the Lepidoptera. Many of these are plant feeders. The order is a very large one, and probably there are many species belonging to it yet undiscovered. A great diversity of habit exists, and in consequence a great variation in the main characteristics.

Mouth Parts.—These are always of a mandibulate type. Well-developed mandibles are nearly always present, though they may not always be used for biting and chewing food (Fig. 13). Frequently they are used for the purpose of carrying food material and building comb. The main modification of the mouth parts tends to take the form of a reduction of the biting parts and a lengthening of the labium into a sucking trunk or *ligula* well seen in most of the bees (Fig. 87). They are thus enabled to extract juices from the nectaries of flowers.

Wings.—The general number of wings is four, and these are well developed and membranous. The fore wings are larger than the hind ones. Frequently the fore and hind wings on either side are coupled together during flight by small frenal hooks situated on the anterior margin of the hind wing. These engage in a frenal fold on the posterior margin of the fore wing (Fig. 10). A few species, chiefly parasites, have lost their wings. The apterous condition is also found in some castes of ants.

Metamorphosis.—This is invariably complex.

Egg.—The general shape of the eggs is ovoid, or rather elongate. In many of the parasitic insects the egg is provided with a stalk or pedicel.

Fig. 87.—Bumble Bee showing elongated mouthparts.

Larva.—Two more or less well-defined types of larvæ are found—those possessing legs and those without. Larvæ of the first type have three pairs of legs on the thoracic segments and six or eight pairs of abdominal pro-legs. The head is always well developed, and these larvæ consequently bear some resemblance to Lepidopterous larvæ. They have always more than eight pairs of legs, however, and thus may be readily distinguished (Fig. 23). Occasionally where larvæ of this type have become wood-borers, all legs are lost except rudiments of those on the thorax. The legless type of larva is very maggot-like in form (Fig. 24, *B*). There is often a conspicuous head present, especially in some of the parasitic larvæ, but in others it is almost invisible.

Pupa.—The pupa is always of the libera or free type, but it is enclosed in a delicate cocoon, which in some cases is made of silky material, and in others of a parchment-like substance. Notably in the bees an intermediate stage exists between the larva and the pupa proper: this stage is known as the *pre-pupa*.

The Adult.—While the largest Hymenoptera are not so big as many other insects, quite a considerable size is attained by bumble bees, hornets and wood-wasps. All sizes, down to the very minute egg parasites, may be found. The commoner insects all have a well-developed head, four gauzy or membranous wings, and usually an ovipositor or sting. The body is more or less heavily chitinized, and often very brightly coloured and hairy. One very noticeable feature, well illustrated by the wasp, is the very thin petiole by which the abdomen is attached to the thorax (Fig. 1). In some of the parasites this is even more pronounced. Other Hymenoptera, however, have no such petiole, and this characteristic difference affords the basis for the first sub-division of the order. It should be observed here that the petiole does not, strictly speaking, join the thorax and abdomen. The first abdominal segment is small and fused to the metathorax, though almost invisible from above: the petiole thus really connects the first and second abdominal segments (Fig. 8).

The main divisions of the order are as follows.:—

HYMENOPTERA (ORDER)

- *Symphyta* (or *Sessiliventres*) (SUB-ORDER)
- *Apocrita* (or *Petiolata*) (SUB-ORDER)
  - Series Aculeata
  - Series Parasitica.

The *Symphyta* have no 'waist,' the abdomen being broadly sessile: the trochanter is two-jointed (Fig. 88), and the larva has both thoracic and abdominal feet.

The *Apocrita* have the abdomen separated from the thorax by a marked construction, the waist or petiole; hence the alternative name *Petiolata*. The trochanter has one or two joints, and the larva is legless.

SYMPHYTA (Sub-Order)

Individuals belonging to this sub-order are considered to be the more primitive forms, on account of their sessile or stalkless abdomens. Their instincts are not highly developed, and their ovipositors are modified into sawing organs.

There is one Super-family, the *Tenthredinoidea*, belonging to which are three families of economic importance:

1. *Cephidæ*—Stem Saw-flies.
2. *Siricidæ* — Wood-wasps.
3. *Tenthredinidæ*—Saw-flies.

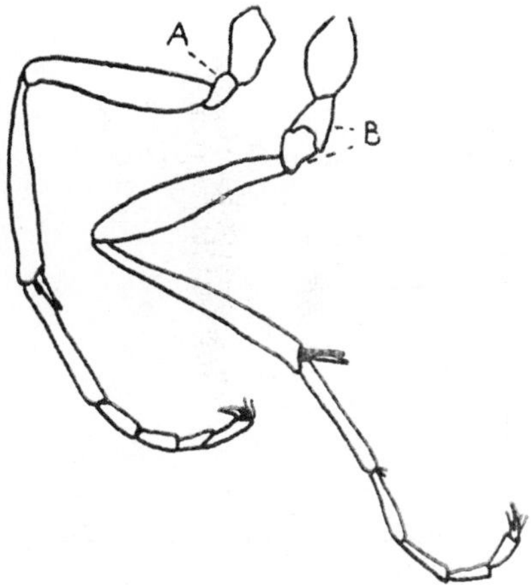

Fig. 88.—Single and Double Trochanters. (A) Leg of Vespid; (B) leg of Ichneumonid. After Buckhurst.

1. Fam. Cephidæ.

Stem Saw-flies.

This is a small family, the members of which may be recognized by the long prothorax, and by the single spine on the front tibia. The larva lives in the stems of grasses, etc., and is legless. There are nine British species. The commonest is

The Corn Saw-fly (*Cephus pygmæus*).

Description and Life History.—The adults are almost one-third of an inch long, and the body is marked with black and yellow bands (Fig. 89). They appear from May to July, when the female flies to the developing wheat plants and deposits a single egg in a slit made with her ovipositor just below the ear. The whitish larva then proceeds to bore up and down inside the stem, until it nearly reaches ground level. It then bites round the inside of the stem and completely severs the straw, causing the crop to 'lodge,' or become laid. The larva hibernates in the stubble, pupating in a slight silken cocoon in spring.

Plants attacked and damage done.—All cereals may be attacked. In this country the pest seldom causes extensive losses, but in Europe and North America it may be very destructive. Attacked plants have thin, whitish-looking ears, with poor, undeveloped grain. At about harvest time the crop may go down badly owing to the biting off of the straw.

FIG. 89.—Corn Saw-fly (*Cephus pygmæus*).

Control Measures.—Burning the stubble after an attack is the obvious way of controlling the pest. On those farms where 'twitch' is regularly collected from stubbles and burned there can be little doubt that many kinds of injurious insects are annually destroyed, including the larvæ of the Corn Saw-fly.

2. FAM. SIRICIDÆ.

Wood-wasps or Horn-tails.

These are perhaps the largest Hymenopterous insects found in the British Isles. Their bodies are long, conspicuous and highly chitinized, that of the female being furnished with a long narrow ovipositor very much like a large sting. Wood-wasps occasionally find their way into houses and cause much alarm on account of their gaudy colouring and long ovipositor. They are frequently mistaken for hornets, but actually they are quite harmless when handled.

The female Wood-wasp lays her eggs singly at the bottom of deep holes drilled in the bark of trees by her ovipositor. The larva which hatches out is whitish, with three pairs of much reduced thoracic legs and no abdominal pro-legs.

A spine on the last segment assists it to move about. It burrows into the heart-wood, and feeds for eighteen months or so before pupating in a cocoon made of silk and small pieces of gnawed wood.

Coniferous trees are usually attacked, but it is very doubtful indeed if wood-wasps can be considered as pests, at any rate in this country. In nearly every case the trees attacked by wood-wasps are already dead or diseased, and the damage caused by the insect is of a secondary nature.

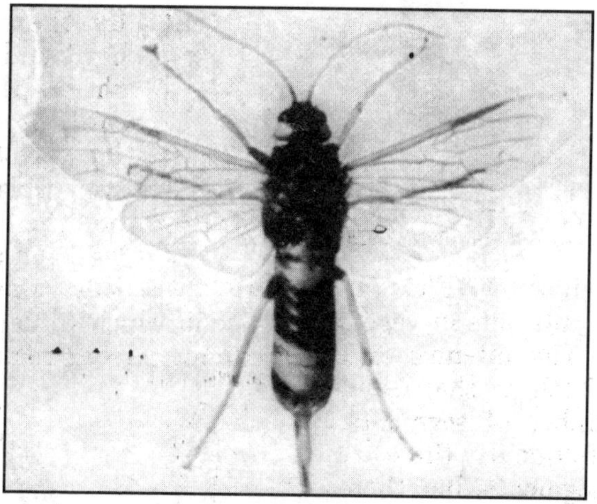

FIG. 90.—Giant Wood-wasp (*Sirex gigas*).

Damage may, however, be caused to felled timber which is left lying on the ground too long, but the advice formerly given to burn affected trees is much too drastic. A large tree is seldom completely ruined by the larvæ, many of which would perish whilst the tree is being sawn up.

The commonest species is the Giant Wood-wasp (*Sirex gigas*) (Fig. 90). It is over an inch long, with a black and yellow body and large tawny wings. The Steel-blue Sirex or Wood-wasp (*S. juvencus*) is smaller and steely blue in colour (Fig. 91). It is less common in this country, but on the Continent it causes considerable losses in pine forests. The Small Blue Sirex (*Hiphydria dromedarius*) occasionally feeds on willows in this country, but does little damage.

FIG. 91.—Steel Blue Wood-wasp (*Sirex juvencus*).

3. FAM. TENTHREDINIDÆ.

Saw-flies.

The Saw-flies may be distinguished from members of the other two families by their broad head and thorax, their plumper-looking bodies, the presence of two tibial spurs on the anterior legs, and the saw-like ovipositor (Fig. 92). With these saws the female insect cuts slits in vegetable tissue in which to deposit her eggs. The antennæ of Tenthredinidæ are extremely variable, both in shape and in number of segments. Most frequently there are nine segments, but there may be as few as three or as many as forty in different species. The adults may be vegetable feeders (phytophagous) or carnivorous. One of the most striking things about saw-flies is the great scarcity of males; in one-third of the recorded British species males are completely unknown. Parthenogenesis is therefore a common method of reproduction.

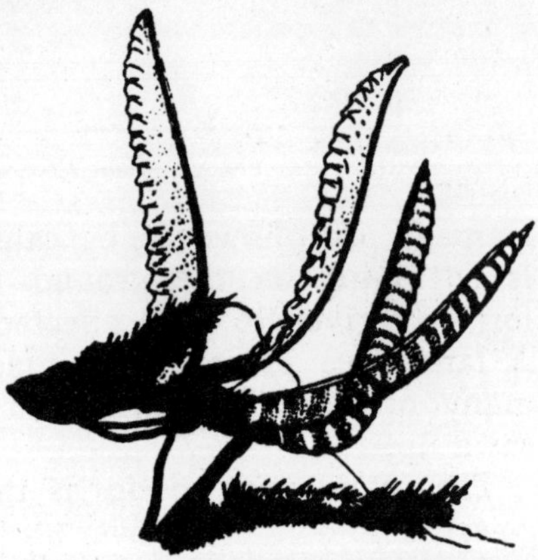

FIG. 92.—Saws of a Saw-fly. (Buckhurst.)

The larvæ of saw-flies are known as 'false caterpillars,'

since they bear a superficial resemblance to lepidopterous larvæ. Saw-fly larvæ may be distinguished from those of the Lepidoptera by the fact that the former have at least six pairs of abdominal pro-legs, while the latter never have more than five pairs (Fig. 23). The head of the larva is hard and chitinized, and provided on either side with a single ocellus. Frequently the body segments are subdivided by a regular number of transverse folds. Occasionally the thoracic segments are larger than the abdominal ones (Fig. 23, *C*), and the body may be covered with a slimy secretion, as in the slug-worms. The larvæ are entirely vegetarian. Some of them cause large galls, such as the bean-galls of the willow. They seem to be distasteful to birds, and make little or no attempt to conceal themselves.

Pupation takes place in the soil either in a silken cocoon or in an earthen cell.

There are over 400 British species, of which very few are of economic importance. The most important are the following :—

The Gooseberry and Currant Saw-fly (*Pteronus = Nematus ribesii*).

Description and Life History.—The female is one-third of an inch long, the head, thorax, and fore part of the abdomen being black and yellow, and the hinder parts of an orange-yellow colour. The first brood appears in April or May. By means of her fine 'saws' the female cuts slits in the veins of the under side of currant and gooseberry leaves and lays her eggs in them. The eggs hatch in a week, giving rise to almost naked larvæ, which are pale green in colour, with black spots over the body. The larvæ of this saw-fly may easily be recognized by the peculiar way in which the last three abdominal segments curve downwards (Fig. 93). There are six pairs of abdominal pro-legs. In about a month, when almost fully grown, the larvæ undergo a remarkable change of colour. The head becomes green, the black spots disappear, and the body turns pale green. There is an orange-coloured area just behind the head and another at the tip of the abdomen. The larvæ make their way into the soil and pupate in small, oval cocoons covered with particles of earth. In less than three weeks the adults emerge; there may be as many as three or four broods per season, the last occurring as late as October.

Plants attacked and damage done.—Gooseberries suffer most from this pest, but red currants and sometimes black currants may also be attacked. In bad cases the entire foliage of the bushes is consumed by the larvæ, with consequent loss of crop and weakening of the bushes.

Natural Enemies.—These saw-flies do not appear to be kept in check to any extent by natural enemies. Several species of ichneumon flies have been recorded as parasitising the larvæ, but they do not seem to be an effective check.

FIG. 93.—Larva of Gooseberry Saw-fly (*Pteronus*=*Nematus ribesii*). (Buckhurst.)

Birds, for some reason, do not find these larvæ very tasty and leave them alone.

Control Measures.—When small, the larvæ usually feed in groups, but later they spread over the whole of the bush, with the result that they suddenly appear to increase in numbers. If hand-picking is practised as soon as the first larvæ are seen it is probable that an expensive spraying campaign will be avoided, but the hand-picking must be done before the young 'caterpillars' begin to separate. Leaves showing the characteristic rows of eggs along the veins underneath should be burnt at once.

In a bad case dusting with such substances as hellebore powder or pyrethrum powder may be employed. A very suitable spray fluid is one of the non-poisonous proprietary insecticides. Most of these compositions contain Derris root extract and are very effective. They are harmless to human beings and may be used up to a week before gathering the fruit. Lead arsenate is effective, but a full month should be allowed between the dates of spraying and picking (see p. 261).

The Apple Saw-fly (*Hoplocampa testudinea*).

Description and Life History.—The adult is a quarter of an inch long and rather more than half an inch across the wings. The apex of the abdomen is broad and square: the head is black and the body reddish yellow. Eggs are laid during the end of April and the beginning of May, but a second brood may occur in the latter part of the year. Each egg is deposited separately, one to each flower, on the receptacle just below the calyx. The larva is creamy white with a brown head, and a brownish plate over the last abdominal segment. Later, this plate turns greyish. The larva has seven pairs of pro-legs and it enters the young fruitlet by boring a hole through the side and not through the 'eye' as in the case of the Codling Moth (see p. 118). It eats out a large one-sided cavity in the fruit, and may migrate therefrom to another apple which is similarly eaten away. Attacked fruit drops before it is ripe. When fully fed the larva leaves the apple, and makes its way into the soil where it pupates in a yellowish, silken cocoon.

Trees attacked and damage done.—Once apples are attacked the interior of the fruit is eaten out and reduced to a wet mass of brownish frass. Frequently there may be seen narrow, ribbon-like scars on the skin of an apple where larvæ have unsuccessfully tried to bite their way in.

Control Measures.—In all cases where young apples drop off the trees a careful examination of them should be made to find out whether the Codling Moth or the Apple Saw-fly is at work. If the latter, it is useless to think of spraying with lead arsenate as recommended for the Codling Moth, because the saw-fly larva enters through the side of the apple, where spray poisons do not adhere very well. Measures against this pest must be directed towards destroying the larvæ either in the apples, or in the soil over winter. As far as possible infested fruits should be destroyed, and the soil round the trees should be stirred up so that poultry may pick out the larvæ and cocoons. Naphthalene and kainit applied to the soil during winter may help to destroy the wintering forms.

The Pear and Cherry Saw-fly (*Eriocampa limacina*).

Description and Life History.—The adult is about one-third of an inch long, and shiny black in colour. It appears in June, laying eggs in slits cut in the under side of leaves. The larva which hatches out is much larger in the thoracic region than at the posterior end. It is almost black in colour, and covered with an offensive slime, from which it derives its name of 'slug-worm' (Fig. 94). It lives on the leaf, eventually changing into an orange-yellow 'caterpillar,' which enters the soil to pupate. There may be two or three broods per year, the larvæ of the last brood hibernating in cocoons of silk, and pupating in spring.

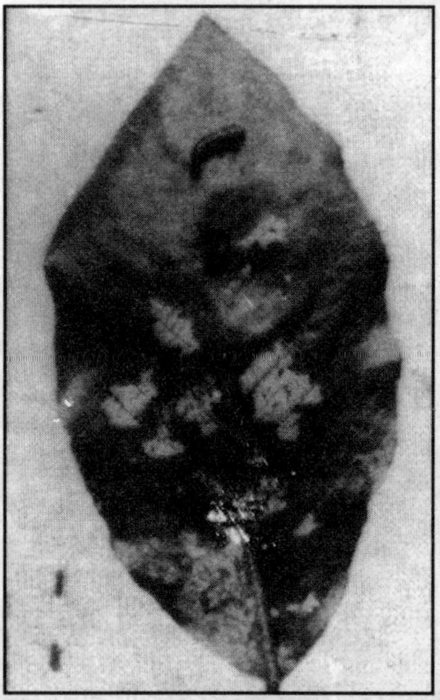

FIG. 94.—Slug-worm or larva of Pear and Cherry Saw-fly.

Trees attacked and damage done.—The pear, cherry and plum may be attacked. The leaf is eaten away on the upper surface, but the lower epidermis is left; thus a brown blotchy appearance is given to the leaf. Serious damage may be caused in severe attacks, although owing to the curious manner in which feeding is carried out the damage is not so conspicuous as actual defoliation.

Control Measures.—Lead arsenate can be used against the earlier broods, and hellebore or nicotine against the later ones. A dusting of freshly slaked lime is very useful if only a few trees have to be treated.

The **Pine Saw-fly** (*Lophyrus pini*) and the **Large Larch Saw-fly** (*Lygæonematus erichsonii*) occasionally do considerable damage to pine and larch trees. The larvæ devour the needles and bark, and may kill young trees. Very little can be done either in the way of prevention or remedy against these pests. Their appearance in large numbers is sudden, and they as suddenly disappear.

APOCRITA (Sub-Order)

The vast majority of Hymenoptera fall within this sub-order, the chief characteristic of which is the 'stalked' or petiolate abdomen. The adults are highly specialized in their habits and are frequently social, *i.e.* they live in communities. The larvæ are legless, and the head may or may not be well developed. They may be phytophagous (vegetable feeders), predaceous, or parasitic.

For convenience we may divide this sub-order into two series, according to the purpose for which the ovipositor is adapted—

Sub-order APOCRITA	Series *Aculeata.*—Ovipositor designed for stinging.
	Series *Parasitica.*—Ovipositor designed for piercing.

Series *Aculeata.*—Stinging Hymenoptera.

Four super-families occur in this series, namely :—

1. *Formicoidea*—Ants.
2. *Apoidea*—Bees.
3. *Sphecoidea*—Digger wasps.
4. *Vespoidea*—Social and Solitary wasps.

Super-Family.—1. *Formicoidea*—**Ants**

There is only one family here, the Formicidæ, members of which can readily be recognized by the appearance of the first abdominal segment. This is always modified into a curious node or scale, leaving the remaining abdominal segments to form an ovoid structure, the *gaster* (Fig. 95). Another distinguishing feature is the geniculate (or elbowed) antennæ.

All British ants are social in habit, living in permanent, underground nests. In the nest may be found winged males (♂), winged females (♀) and wingless workers (☿). These last are undeveloped females. The female usually, but not always, possesses a sting. What is commonly called an ant's 'sting' is really a bite inflicted by a worker-ant with its powerful mandibles, after which a small amount of formic acid is squirted into the wound. Winged males and females are produced during summer: these leave the nest and pair in the air. The female on reaching the ground sheds her wings, and lays a few eggs in a suitable site. These eventually develop into workers, and the 'queen,' relieved of the duty of looking after developing larvæ, devotes herself entirely to egg-laying. Ants' eggs are very minute, seldom more than .5 mm. long. The commodity sold in shops as 'ants' eggs' is either the larvæ, pupæ, or cocoons of ants.

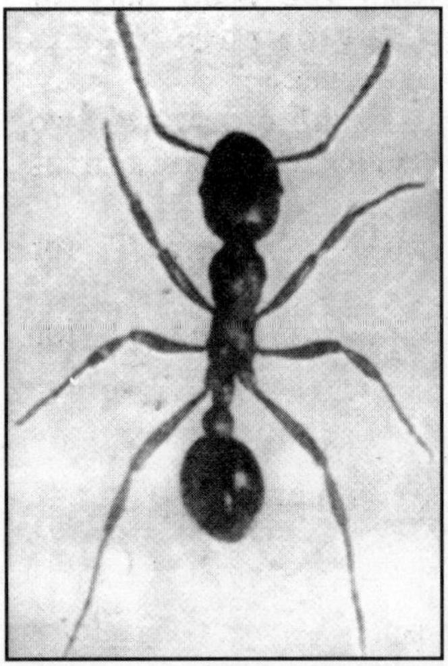

FIG. 95.—A Common Ant (*Myrmica ruginodis*), showing the shape of the abdomen.

Ants are not of very great agricultural importance, though they are great scavengers and do a lot of good in clearing up carrion, dead insects, etc. Occasionally they may become a serious nuisance in pastures or meadows since their large nests get in the way of implements, particularly the mowing machine. Ants' nests may be destroyed with a spade, the ant-hill being broken up and the soil dispersed, or a few holes may be cut in the top layers and about an ounce of carbon disulphide poured in. The holes should be blocked up with a thick turf. A soil injector can be used for this purpose as in the case of root attacks by Woolly Aphis. Some species of ants feed chiefly on the sweet liquids excreted by aphides and scale insects. These insects are tended by the ants, which take them into shelter during bad weather and carry them back to food places when conditions are favourable. The ants, therefore, become harmful insects inasmuch as they foster the plant-feeding aphides.

SUPER-FAMILY.—2. *Apoidea*—**Bees**

All members of this super-family have more or less hairy bodies, and the hairs are characteristically branched, or plumose. The femora, tibiæ or tarsi on the hind legs are modified into 'pollen baskets.' These areas, often enlarged, are covered with hairs to retain the pollen grains which are used as food.

There are more than 200 British species, spread over numerous families. Included here are the Leaf Cutting, Mason, Carpenter, Cuckoo and Bumble bees, as well as the social Honey-bee (*Apis mellifica*). Bees do an enormous amount of good in cross-pollinating flowers. The literature dealing with them is very vast, and the student is referred to the books mentioned in the Appendix for details of their habits, etc.

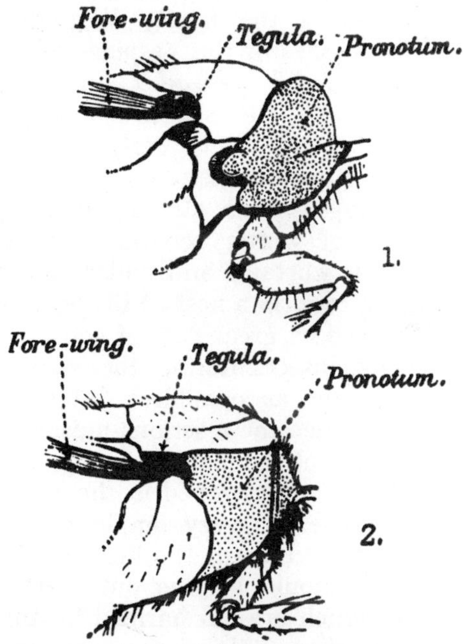

FIG. 96.—1. Thorax of *Sphecoidea* (Pronotum not reaching Tegula.) 2. Thorax of *Vespoidea* (Pronotum reaching Tegula). (Buckhurst.)

SUPER-FAMILY.—3. *Sphecoidea*—**Fossorial Wasps**

The fossorial or digging wasps differ from the true wasps in that the pronotum does not extend back to the tegulæ as it does in the Vespoidea (Fig. 96). The Sphecoidea have simple, unbranched body hairs. They make their cells in the ground and store them with various insects or larvæ which they paralyse with their stings. They may be regarded as a beneficial family.

SUPER-FAMILY.—4. *Vespoidea*—**Social and Solitary Wasps**

The wasps have the pronotum extending back to the tegulæ (Fig. 96), whilst the trochanter is almost always one-jointed (Fig. 88). The body hairs are unbranched. A large number of families is included here, of which the most important is the family Vespidæ or social wasps. All British species of the Vespidæ belong to the genus *Vespa*. *V. vulgaris*, *V. germanica*, and *V. rufa* make their nests underground; *V. sylvestris* and *V. norvegica* hang their nests from trees; *V. crabro* nests in hollow trees. The nests endure for one season only. With the exception of the young queens all the wasps perish when the cold weather of winter sets in. The queens find shelter in houses, in curtains and under pictures, or elsewhere, commencing to form a nest at the beginning of May.

Wasps do a great deal of good by destroying all sorts of injurious insects during the summer and should be left alone as far as possible. Later in the season, however, they may become a serious nuisance in orchards, attacking pears and plums and reducing their value. Their nests must then be destroyed in the evening when all the wasps are at home. A very simple way is to pour about a pint of paraffin oil down the hole and block it up. Next day the nest should be dug out and burnt to kill the grubs inside, which are not harmed by the fumes of the paraffin. A lump of cyanide of potassium about half the size of a pigeon's egg thrust down the hole, or a piece of cotton-wool soaked in a strong solution of the cyanide, will also kill the adult wasps.

Series *Parasitica*—Parasitic Hymenoptera.

These insects have their ovipositor designed for piercing, not for stinging as in the Aculeata. There are four main super-families :—

5. *Proctotrypoidea* (or *Serphoidea*).
6. *Cynipoidea*—Gall flies.
7. *Ichneumonoidea*—Ichneumon flies, etc.
8. *Chalcidoidea*—Chalcid flies.

SUPER-FAMILY.—5. *Proctotrypoidea* (*Serphoidea*)

Most of these insects are very small. The pronotum extends backwards to the tegulæ, and the trochanters are usually two-jointed. The ovipositor is in the form of a cylindrical tube and arises fróm the tip of the abdomen.

A number of species are egg parasites, while others are parasitic in the bodies of other larvæ. A few are *hyper-parasites* (Fig. 97). A silken or parchment-like cocoon is usually formed by the larvæ, and within this pupation takes place. In those species which parasitise aphides, the

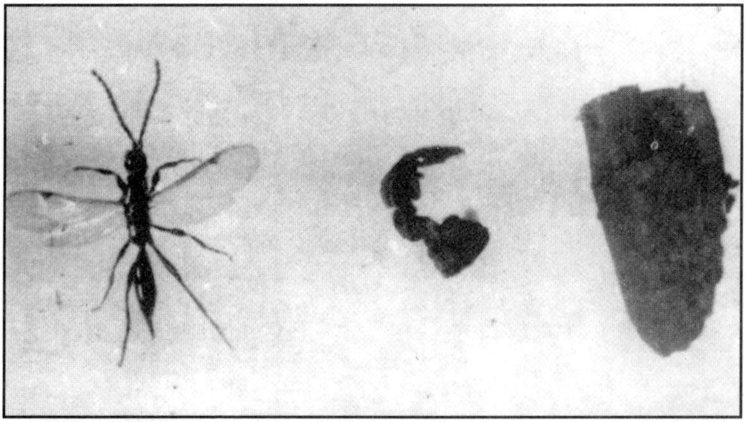

FIG. 97.—*Exallonyx ligatus*, a Proctrotrypid hyper-parasite. It was bred from the larva of a beetle (centre) which preys upon the Cabbage Root-fly, whose pupal case is seen on the right. (K. M. Smith.)

larvæ generally pupate within the dried-up bodies of their hosts. One of the best known species is *Platygater minutus*, which is a parasite of the Hessian fly.

SUPER-FAMILY.—6. *Cynipoidea*—**Gall Flies or Gall Wasps**

These insects may be recognized by their laterally compressed bodies, and by the absence of a stigma on the fore wings. The trochanter is undivided, the antennæ are not elbowed, and consist of not more than fifteen joints. There are two families, Cynipidæ and Figitidæ, of which the first is the larger and better known. Members of the Cynipidæ are called gall flies or gall wasps, on account of the property that many possess of causing gall-like swellings in vegetable tissues. All Cynipidæ, however, do not form galls. A few species parasitise flies, aphides and other insects. *Cothonaspis rapæ* (Fig. 98) is a common parasite of the Cabbage Root Fly. Others (Synerginæ) do not themselves form galls, but live in the galls produced by other species. Such species are called *inquilines* or 'guests.' In many of the gall wasps, males have never been discovered. In others a remarkable alternation of generations has been observed. The overwintering forms are all parthenogenetic females. Reproduction takes place without the appearance of the male until the summer, when both males and females are produced. The galls formed by these two generations may be very different, and the two generations have frequently been described as different species. It has been stated that 86 per cent. of the known species of gall wasps produce galls on the oak, 7 per cent. on the rose, and the remaining species on other angiosperms, particularly Compositæ.

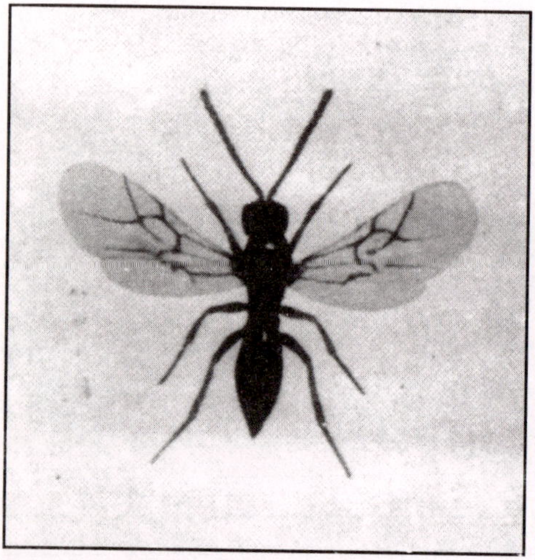

FIG. 98.—A Cynipid (*Cothonaspis rapæ*), a parasite of the Cabbage Root-fly. (K. M. Smith.)

One kind of 'oak-apple' is caused by *Biorrhiza pallida*. On cutting open an oak-apple gall in early summer the whitish apodous larvæ may be discovered. These develop in July into winged males or wingless females. The apterous female crawls down the trunk of the tree and on to the roots, and proceeds to lay eggs within the root tissue. As a result, galls are formed, and from each one there comes in spring another apterous female, which crawls up the tree and lays eggs parthenogenetically in the terminal buds.

The hard 'marble galls' of the oak are caused by a different species, *Cynips kollari*.

SUPER-FAMILY.—7. *Ichneumonoidea*—**Ichneumon Flies, etc.**

The members of this group agree in having non-elbowed antennæ and two-jointed trochanters. The pronotum extends to the tegulæ, and the ovipositor issues some distance before the tip of the abdomen. There are three families, Evaniidæ, Ichneumonidæ and Braconidæ, members of the first-named being somewhat rare. The Ichneumonidæ and Braconidæ are very much alike in structure and appearance, but they may be distinguished by the venation of the fore wing. In the Ichneumonidæ there is a large single cell in the centre of the wing formed by the fusion of two smaller ones (1st discoidal and 1st cubital). Ichneumon 'flies' may be readily recognized by their curved, sickle-shaped abdomens; also by their quivering antennæ, which move alternately up and down whilst the insects are seeking a place for egg-laying (Fig. 99).

FIG. 99.—An Ichneumon (*Pimpla*).

The habits of the two families are also very similar, and may be considered together. They are all parasitic, chiefly upon larvæ of Lepidoptera, Hymenoptera and Coleoptera, whilst certain kinds of Braconidæ specialize upon aphides. Some are egg parasites. The females frequently hibernate, but the males die off before winter. Apterous females are common in both families. The adults are on the wing on bright, sunny days, and may be found in numbers upon umbelliferous plants: some species (*Ophion*) fly at night and are attracted by artificial light. They are large, reddish, long-legged 'flies,' which parasitise noctuid larvæ.

The females are provided with long ovipositors, with which they lay their eggs in the body tissues of the larva to be parasitised. The egg hatches into a tiny legless grub, consisting of a head and thirteen other segments. This feeds upon the fat body of the host and other organs, carefully avoiding the vital regions, injury to which would cause the death of the host and consequently of the parasite as well. When full-fed the larva pupates either within the host or outside it, forming a yellow, or black and white, cocoon of silk.

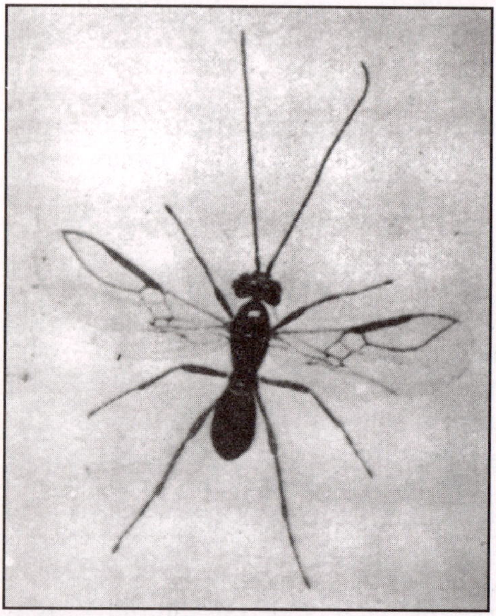

FIG. 100.—A Braconid (*Dacnusa stramineipes*). (K. M. Smith.)

Apanteles glomeratus is a Braconid which severely parasitises the larvæ of the Cabbage White Butterflies (*Pieris spp.*). As many as 150 grubs of this species may emerge from a single caterpillar, which becomes covered or surrounded with the bright yellow cocoons of the parasite.

Economically the Ichneumonidæ and Braconidæ are of very great importance, as they act as natural checks upon many of the worst insect foes of the farmer and gardener. They should be encouraged by every possible means. Recently, certain species have been introduced from one country to another in an attempt to cope with pests which are difficult to control by artificial means (see p. 256).

SUPER-FAMILY.—8. *Chalcidoidea*—**Chalcid Flies**

Chalcids are small insects with sharply elbowed antennæ and much-reduced wing venation: frequently they have bright metallic colourings. There are over 1400 British species, spread over numerous families. Some species live in seeds or stems, but the majority are parasitic. Some are hyper-parasites, that is, they parasitise the parasites of other insects. The eggs and larvæ of Lepidoptera are commonly attacked, whilst certain kinds of chalcids cause great destruction amongst Coccidæ. *Nasonia brevicornis* parasitises the pupa of the Blow fly, and *Aphelinus mali* has been introduced into England in an attempt to control Woolly Aphis.

CHAPTER XVI

DIPTERA—TRUE FLIES

MANY insects which do not belong to this family go by the name of 'flies,' largely because they possess clear membranous wings and have considerable powers of flight. The name Diptera should, however, be strictly confined to the true flies, which differ from all other common insects in possessing two wings only.

The Mouthparts of Diptera exhibit a good deal of variation. The commonest type is that adapted for taking up liquid food only, such as the nectar of flowers, or substances which have been dissolved by the saliva which the fly pours out over its food material. This sucking type, illustrated in Fig. 101, is complicated in structure and very highly specialized. There are many flies, on the other hand, which are blood-suckers. Their mouthparts are adapted for piercing the skin of animals as well as for sucking up the blood which issues from the wound (Fig. 15). The blood-sucking habit is confined to the female insect in some families, but in others both sexes feed in this way. In a few families the extreme case is met with, in which the mouthparts are entirely undeveloped and the adult insects appear never to feed at all.

Wings.—The true flies never possess more than two wings and these are invariably found on the mesothorax. They are membranous in texture and are never hardened nor covered with scales, though they are sometimes coloured by hairs or scales arranged in bands or in patches. On the metathorax a pair of small organs, known as *halteres* or *balancers*, are present in place of wings. These take the form of a short stem carrying a knob at the end, similar to a tiny drumstick (Figs. 105 *B*, 104). In a few specialized families of flies the wings are entirely absent, due to the parasitic mode of life taken up by the insects.

Metamorphosis.—All flies exhibit a complex metamorphosis, the four stages, egg, larva, pupa and imago, being clearly defined. Exceptional cases occur, however, in some of the flies parasitic upon animals (Pupipara), where

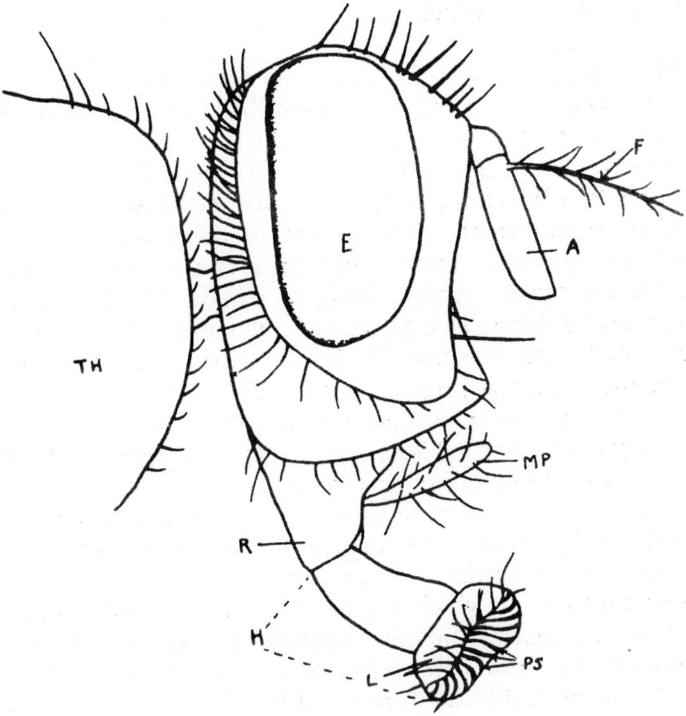

FIG. 101.—Head of a Muscid-fly showing sucking mouthparts (diagrammatic).

A. antenna : *E.* eye : *F.* flagellum : *H.* haustorium : *L.* labellum : *MP.* maxillary palp : *Ps.* pseudo-tracheæ : *R.* rostrum : *TH.* thorax.

the egg hatches and the larva develops within the body of the female. The female therefore 'lays' a fully grown larva, which pupates almost at once.

The Egg.—In the majority of cases, the egg remains as such for a very short period. It is thin-shelled as compared with some of the eggs of Rhynchota and Lepidoptera, and hatches quickly. The shell is frequently very delicately sculptured. It is white in colour in most cases, but dark-coloured eggs are the rule in certain families. Many flies, whose larvæ are aquatic in habit (Culicidæ), lay their eggs in masses on the water. The eggs float about until they hatch, and the larvæ escape from a hole in the bottom into the water. Some very remarkable eggs are those laid on the hairs of animals. Such eggs have a short stalk provided with two flanges, which clasp round the hairs and fix the egg firmly to it.

The Larva.—Dipterous larvæ never possess legs. They move about by means of worm-like contractions and expansions of the body, assisted by tubercles or swellings on the ventral side. These tubercles are called false legs or *pseudopods*, and are much more developed in some families than others. The degree of development of the head is very varied, and the mouthparts are often of a very specialized type. Only vestiges remain of the normal parts, and feeding is carried out by means of a scraping movement with the mandibular sclerites. In such larvæ no definite head is present at all. The anterior segments contain a set of dark chitinous plates known as the *cephalo-pharyngeal skeleton*. These plates are in three different parts, a pair of mandibular sclerites or mouth-hooks at the anterior end, an intermediate sclerite, and a pharyngeal sclerite (Fig. 102). All degrees of development occur beyond this condition. Some larvæ possess a so-called 'head capsule' in which the head end of the body is more or less chitinized, but the capsule is capable of being retracted into the first segment of the thorax so as to be almost invisible at times. Other larvæ have a fully developed head and mandibulate mouthparts. The breathing system is usually very much reduced and many pairs of spiracles are permanently closed. In most newly-hatched larvæ only one pair of spiracles is open, namely that on the last abdominal segment. This arrangement is also found in the adult larvæ of the Crane flies (Tipulidæ) and the Ox Warble fly (*Hypoderma*). Two main tracheal tubes run into the body from these spiracles. The most usual arrangement is for two pairs of spiracles to function, the prothoracic pair and the pair on the last abdominal segment. Most of the larvæ of flies in the sub-order Cyclorrapha exhibit this arrangement. The prothoracic

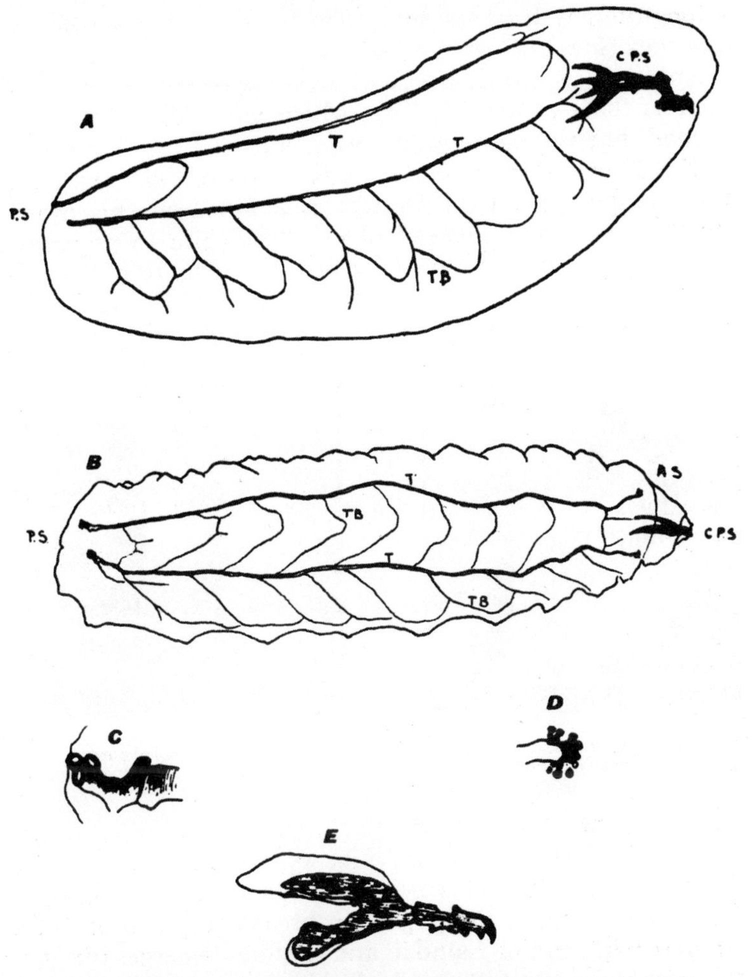

Fig. 102.—Larvæ of Diptera, showing breathing system.

A. newly hatched larva of mangold fly much enlarged. *B*. fully grown larva. *C*. posterior spiracle (enlarged) of *B*. *D*. anterior spiracle of *B*. *E*. cephalo-pharyngeal skeleton. *PS*. posterior spiracle: *AS*. anterior spiracle: *T*. trachea: *TB*. tracheal branches: *CPS*. cephalo-pharyngeal sclerites.

Note.—*A*. is drawn to a larger scale than *B*.

spiracles project slightly from the surface of the body, and are often branched or fan-shaped. Each prothoracic spiracle is connected to the corresponding posterior one by a longitudinal tracheal tube, which gives off branches at intervals (Fig. 102).

Antennæ, as a rule, are poorly developed, except in larvæ such as those of the Tipulidæ, which move about in search of food, but they are never conspicuous organs.

The Pupa.—Two definite kinds of pupæ exist, and the main subdivision of the Diptera is founded on this difference. Some larvæ, instead of moulting prior to pupation, retain their skin and pupate inside it. This skin becomes barrel-shaped and hard in texture due to chitin laid down

FIG. 103A.—Larva and pupa of Orthorrapha (Crane-fly).

FIG. 103B.—Larva and pupa of Cyclorrapha (Calliphora).

on it. It forms a protective covering for the pupa inside, and is called the puparium (Fig. 25, *B*). The pupa itself is white and soft. This type of pupa is called coarctate. The anterior end of the puparium breaks off by means of a circular split running round it, and the adult emerges through the opening. The flies in which this occurs are classified as the **Cyclorrapha** (=circular split). In the other type the skin is moulted normally prior to the formation of the pupa, which is itself covered with a chitinized skin. A longitudinal or T-shaped split in this skin allows the adult to emerge. Flies exhibiting this characteristic are known as the **Orthorrapha** =straight split) (Figs. 103A, 103B).

The Adult or Imago.—The main division of the Diptera into Cyclorrapha and Orthorrapha is based on pupal characters. Further subdivisions, however, are made on adult characters.

The head is relatively large in all flies and capable of considerable movement. The compound eyes are often of great size and almost meet in the median line, especially in males. They are then said to be *holoptic*. Three ocelli are found on the vertex in the form of a triangle. The antennæ may be long and threadlike, composed of a large number of segments, or short, with only three segments, the terminal one of which is often a bristle or *arista*. In some flies (Schizophora) a small crescent-shaped scar may be seen just above the antennæ (Fig. 104). This is the *frontal lunule*, and from it protrudes a sac-like structure, the *ptilinum*, when the fly emerges from the puparium. This organ is of service in pushing off the cap from the puparium, and once the fly has emerged, it is withdrawn into the head.

A

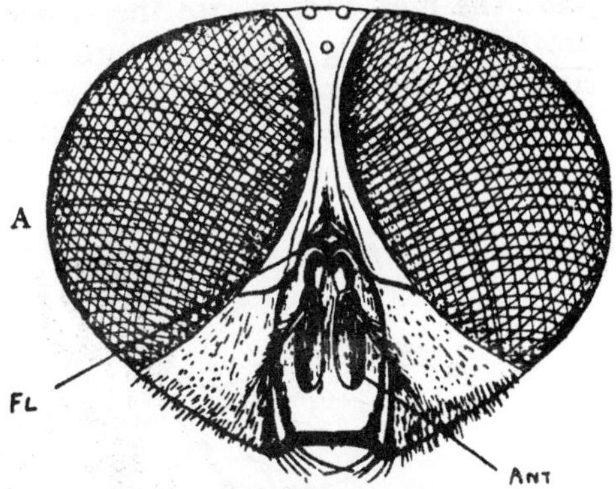

FIG. 104A.—Head of Schizophora showing frontal lunule or cavity.
FL. Frontal lunule; *Ant.* Antenna.

B

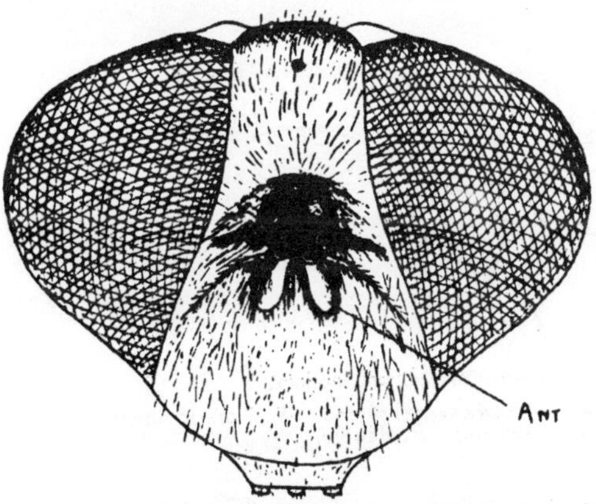

FIG. 104B.—Head of Aschiza (frontal lunule absent).
Ant. Antenna.

The thorax shows a great development of the middle segment. The prothorax and metathorax are greatly reduced in size, while the mesothorax, which carries the wings, is large. On the hind margin of the wing, where it joins the mesothorax, are the *squamæ* or *calypteræ*. These are lobes arising from the wing and they vary greatly in size. In the Acalypteræ they are small and do not cover the halteres, while in the Calypteræ they are very large

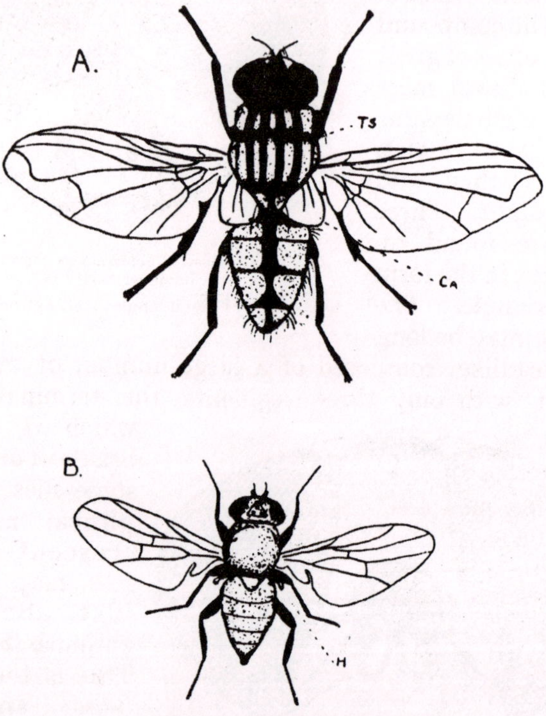

FIG. 105.—Typical Schizophora.
A. Calypteræ (House-fly); *B. Acalypteræ* (Frit-fly): *TS.* Thoracic suture: *Ca.* Calypteron: *H.* Haltere.

and the halteres are concealed by them (Fig. 105). The abdomen possesses no unusual characteristics. An ovipositor is usually present, but it is often retractile and not in evidence except when eggs are being laid.

The whole body of the fly is soft and there is no heavy protective chitin covering. It is frequently very hairy or scaly, and its colours are often due to these scales and hairs.

The order may be classified as follows :—

DIPTERA (ORDER)

- *Orthorrapha* (SUB-ORDER)
  - SERIES: *Nematocera* (*Nemocera*)
  - *Brachycera*
- *Cyclorrapha* (SUB-ORDER)
  - *Athericera*
    - SECTION: *Aschiza*
    - *Schizophora*
      - SUPER-FAMILY: *Acalypteræ*
      - *Calypteræ*
  - *Pupipara*

SUB-ORDER.—1. ORTHORRAPHA.—This sub-order includes those flies which emerge from the pupa through a longitudinal slit. The two series into which it is further divided have the following characteristics :—

SERIES 1. *Nematocera.*—Larva with a definite head, the mandibles of which work in a horizontal plane. The pupa is of the 'libera' type. The name Nematocera, which means 'thread-like antennæ,' refers to the long many-jointed antennæ found in these insects.

SERIES 2. *Brachycera.*—As the name indicates, these flies are characterized by short antennæ, usually of three segments, the terminal one of which is often bristle-like. The larvæ have a reduced type of head which is capable of being retracted into the thorax, and the mandibles work in a vertical plane. The pupa is of the 'libera' type.

SUB-ORDER.—2. CYCLORRAPHA.—The larvæ of these flies form a puparium, the pupa being of the type known as 'coarctate': emergence takes place through the anterior end of the puparium breaking off as a circular cap.

SERIES 1. *Athericera.*—By far the majority of the Cyclorrapha come under this series and further subdivision is necessary.

SECTION A. *Aschiza.*—The frontal suture is indefinite and no trace of the ptilinum is found after the emergence of the adult.

SECTION B. *Schizophora.*—The frontal suture is definite (Fig. 104A), and the ptilinum may be seen after the adult emerges, though it is eventually withdrawn into the head.

SUPER-FAMILY 1. *Acalypteræ.*—Squamæ (or calypteræ) small and not concealing the halteres, and the transverse suture on the thorax is indistinct (Fig. 105, *B*).

SUPER-FAMILY 2. *Calypteræ.*—Squamæ large and concealing the halteres. Transverse suture on thorax well defined (Fig. 105, *A*).

SERIES 2. *Pupipara.*—Adults are all ectoparasites on warm-blooded vertebrates or bees. Their bodies are usually very flattened and hard in texture. Wings may or may not be present. Eggs are not laid, but living larvæ are produced. The pupa is coarctate.

1. ORTHORRAPHA (SUB-ORDER)

SERIES 1. *Nematocera* (*Nemocera*).—The main families which contain species of economic importance are as follows :—

Tipulidæ.—Crane Flies or Daddy-long-legs.
Culicidæ.—Gnats and Mosquitoes.
Cecidomyidæ.—Gall Midges.

FAM. TIPULIDÆ.—Many species belonging to this family are found in England, and their larvæ are the 'leather-jackets' which are often the cause of serious loss to crops. Our most common members are *Tipula oleracea*, *T. paludosa* and *Pachyrrhina spp.*: for all practical purposes these may be considered together.

Description and Life History.—The adults are large flies with a big expanse of wings, almost 2 inches from tip to tip, and possessing long fragile legs. The end of the abdomen in the male is blunt, whilst that of the female is pointed. They are to be found throughout the summer months particularly in pastures or other grassy situations, where they deposit their eggs. These eggs, which are oval in shape and black in colour, are usually deposited while the insect is in flight, and one may easily observe the short jumping flights quite close to the ground made by the female for this purpose. Coarse, tufted grassland is chosen

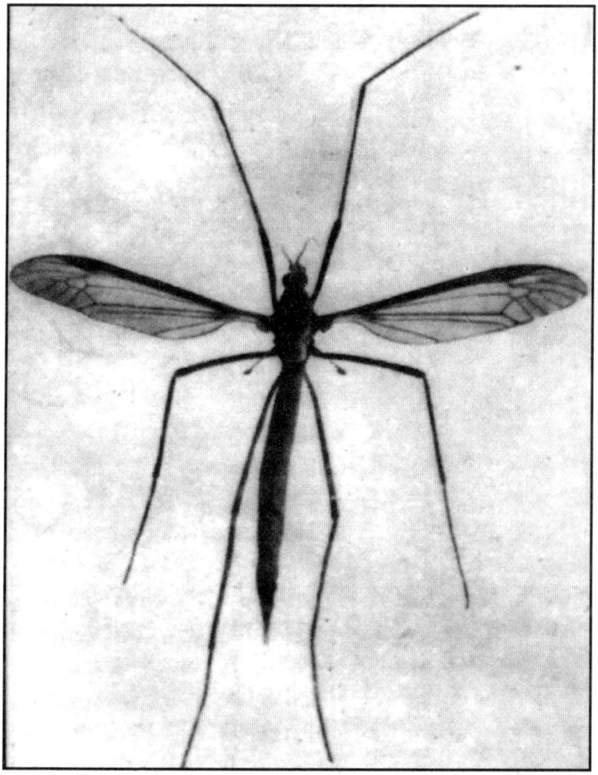

FIG. 106.—Female Crane Fly or Daddy-long-legs (*Tipula oleracea*).

for oviposition, since it affords protection from direct sunlight which, combined with dry soil conditions, is fatal to young larvæ. The eggs hatch in about fourteen days and the young larvæ immediately take shelter underground. The length of time occupied by the larval stage varies with the amount of food which can be obtained, and possibly on other conditions, but normally the larva remains as such for a period of approximately nine months. When fully fed it pupates in the soil. The pupa has considerable power of movement, wriggling its way to the surface so that the anterior end is free before the adult emerges. Owing to the variation in the length of time occupied by the larva in its development, adults emerge over a long period, which lasts most of the summer, and egg-laying is also in progress for this length of time. The number of eggs laid by one female may be in the region of 300, and in an attacked crop upwards of 20 larvæ per square yard may be found.

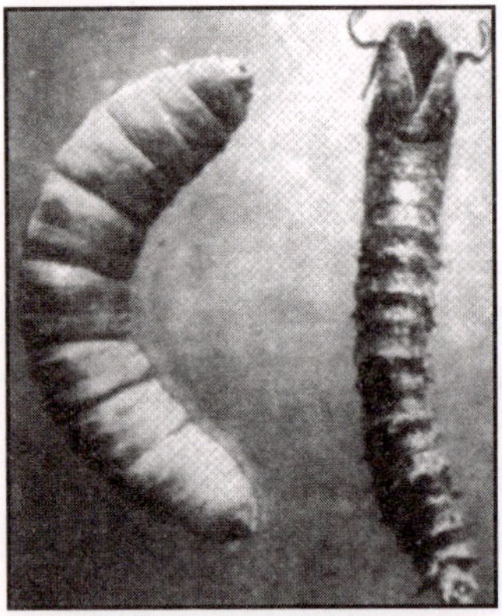

Fig. 107.—Larva and pupa skin of Crane Fly.

The larvæ when almost fully grown are earthen-coloured, legless grubs, with a small dark head, and they are often $1\frac{1}{2}$ inches long. On the tail end two spiracles may be seen, surrounded by fleshy tubercles. Normally they live in burrows just beneath the surface of the soil, but in dull, damp weather they may be found on the surface during the day or under clods of earth and stones. They appear to come to the surface much more at night, however, a point which should be noted in view of control methods. The pupa is somewhat horny in texture, but capable of a certain amount of movement, and it possesses two horn-like projections on the head end (Fig. 107). Only one generation occurs in the year.

Plants attacked and damage done.—The usual food of Crane fly larvæ consists of grass, both roots and blades, but a great variety of plants may be attacked, including tuberous and fleshy portions, into which the larvæ bore. Grassland itself frequently suffers a good deal, large, thin or bare patches occurring where the attack is worst, but, owing to its power of recovery, grass suffers less than many crops. Particularly on golf greens destruction may be very obvious. It is to crops which are sown following on the ploughing-up of grass that the greatest damage is done. Eggs are not laid unless shade and damp conditions can be secured. Weedy stubbles or a long growth of grass on leys during the late summer are therefore liable to lead to an attack on the succeeding crop. Spring-sown cereals following on the ploughing-up of a ley (or of permanent pasture) are most commonly attacked. Even before the seed germinates the larvæ eat it, and from that time onward until they pupate the 'leather-jackets' do more and more damage. The root system of the plant is severely attacked, and at night when they come above ground the larvæ eat the green shoots so that less and less 'plant' shows on the field. In this way the crop may be destroyed altogether in patches and often has to be ploughed up. April and May are the months when such damage is most apparent, the larvæ then being almost fully grown.

Natural Enemies.—Birds are of great assistance in keeping down the numbers of leather-jackets, rooks, lapwings, and starlings feeding readily upon them. Numbers are destroyed in this way when exposed by cultivation. Little good can be done by birds, however, in a growing crop.

Control Measures.

Grassland.—The control of leather-jackets in grassland is largely associated with problems of general management, rather than taking specific action against the larvæ when they are already doing damage. As far as possible all coarse, long, or tufted grass should be removed by late summer so as to discourage the flies from laying eggs there. This may be achieved either by feeding the grass off, or by the use of a mowing-machine, or even a scythe, in bad patches. Such treatment is not only of service in discouraging oviposition by the flies, but is recognized as being good grassland management. It has been suggested that bush harrowing at such times, or rolling heavily when the pupæ are projecting from the ground prior to the emergence of the adult, will do good, but it is difficult to visualize these operations ever being carried out in the ordinary course of farm routine work. A dressing of Naphthalene, preferably in a crude form, has frequently been found beneficial, 2 cwts. per acre being a normal dressing, but on golf courses up to 5 cwts. has been employed. Naphthalene may be sown broadcast, and its effect is most noticeable if considerable rainfall follows on its application, large numbers of larvæ then being killed.

Arable Crops.—It is on ploughed-up grass and clover leys that most damage is done, and the chance of attack is greatly increased if the late summer and autumn have been wet. A cereal crop usually follows a ley in the rotation, and may be either winter or spring sown. In any case it is advisable to get the ploughing done as early as possible, not later than the end of August. If this is impracticable the herbage should be kept short until ploughing can be done. These precautions having been taken, it is important to consolidate the land as much as conditions will allow and, especially with spring-sown cereals, to make a good seed-bed, so that once germination takes place the plants rapidly get a firm root-hold and continue to grow unchecked. In order to assist this a top dressing with nitrate of soda may be worth while, but in spite of top dressings, if large numbers of leather-jackets are present, immense damage will be done.

A poison-bait method of dealing with attacks has recently been found very effective, but the way in which it is at present applied makes its use illegal under the Protection of Animals Act, 1911. This Act says that a person is liable to a penalty if he "knowingly puts or places or causes or procures any person to put or place, or knowingly is a party to the putting or placing in or upon any land or building any poison or any fluid or edible matter (not being sown seed or grain) which has been rendered poisonous." The poison used for leather-jackets is an arsenical compound known as Paris Green, which is mixed with bran and scattered thinly over the field either broadcast or by a drill. Leather-jackets will readily eat the bran at night when they come to the surface, and are thus poisoned. The danger to birds is apparently very small, since the bran rapidly disappears, and given two rainless mild nights the majority of the leather-jackets can be killed. There has been at present no legal action arising from the use of Paris Green, and whether or not it may be considered a justifiable act in agriculture to use it, is still open. A formula involving the use of Paris Green is given in Chapter XIX.

In crops other than cereals little damage is done as a rule, since these crops are sown later than cereals and the leather-jackets are by that time mostly in the pupal state. Frequent hoeings which expose the larvæ to birds are most beneficial if such cases occur.

Occasionally garden crops suffer some damage, particularly on freshly broken-up land, and the same procedure may be adopted among such crops as that outlined above. The larvæ may be trapped under sods laid on the surface of the ground. They will congregate during the day under such shelter and may then be destroyed.

FAM. CULICIDÆ.

Gnats and Mosquitoes.

Although these insects have no agricultural importance they are well known to be carriers of certain disease-producing organisms affecting man, and must therefore receive some mention.

The number of different species found in this country is comparatively small, the majority occurring in warmer countries.

In appearance they are slender, long-legged flies, and possess elaborate piercing mouthparts (Fig. 15), the bite of which is very painful. They suck the blood of various animals and birds, and through the puncture which they make, inject a salivary secretion which prevents coagulation of the blood. They may thus carry the tiny disease-producing organisms from an infected to a healthy person. Such diseases as malaria and yellow fever are well known to be transmitted in this way, and extreme precautions have to be taken in some cases to control the numbers of mosquitoes in order to keep these diseases under control. A well-known example of this is the case of the Panama Canal, where enormous numbers of men engaged in cutting it died from fever, and the project was only carried through when means of controlling the mosquito had been discovered.

The Mosquitoes (*Anopheles*) can be distinguished from

the Gnats (*Culex*) when at rest, as shown in Fig. 108. *Anopheles* rests with the proboscis and body all in a straight line at an angle to the surface on which it is resting. In *Culex* there is no such straight line, the body being reflexed toward the surface on which the insect is resting.

The larvæ of the Culicidæ are aquatic in habit. Eggs are laid on the surface of water, frequently in masses, and the larvæ on hatching swim about, coming to the surface to breathe. Respiration is carried out through a pair of abdominal spiracles which are brought in contact with the surface film. Advantage was taken of this fact in controlling mosquitoes, for it was found that a very thin film of oil on the surface of the water prevented respiration taking place, and the larvæ were killed. Stagnant, rain-filled pools, or even water in tins, etc., form breeding-places for many species. Larvæ are seldom prevalent in permanent waters owing to the fact that they are readily devoured by fish and other organisms.

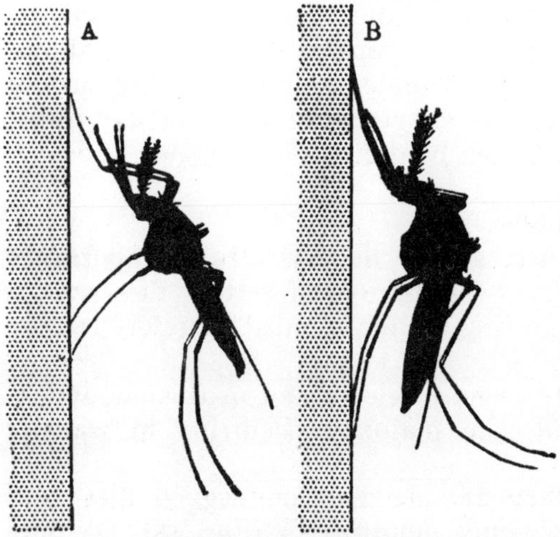

FIG. 108.—Mosquitoes in resting attitude. (British Museum Guide.)
A. *Anopheles*; B. *Culex*.

The pupa is also aquatic and very active, coming to the surface before the adult emerges.

FAM. CECIDOMYIDÆ.

Gall Midges.

Although these flies have received the popular name of Gall Midges, by no means all of them form galls on plants, a number being more or less parasitic or predaceous, while others live on decaying animal or vegetable refuse.

They are all small, delicate, long-legged flies, with whorls of fine hairs on the antennæ and with similar hairs on the wings and body. The larvæ possess no definite head. They may usually be recognized by the possession of a *spatula* or ' breast bone,' a dark chitinous plate which is seen on the sternum of the first thoracic segment. The function of this organ is still uncertain. Pupation may take place inside a kind of cocoon or, in some cases, a covering very similar in appearance to a puparium is formed from the larval skin, which is not completely moulted off. The true puparium is, of course, typical of the Cyclorrapha. Cecidomyidæ may be recognized in flight by their delicate trailing legs.

The Hessian Fly (*Mayetiola = Cecidomyia destructor*).

Description and Life History.—This insect received the name of Hessian fly in America, to which it appears to have been introduced by Hessian troops, having been carried over in straw in the pupal or ' flax seed ' stage, which will be described later. It is about one-tenth of an inch in length, of a smoky black colour, with the abdomen reddish in live specimens. The wings, antennæ and legs are also dusky owing to their covering of fine hairs or scales. The male is distinctly more slender in form than the female (Fig. 109).

The eggs are laid on cereals and grasses and are very small and reddish in colour. They are usually deposited on the upper surface of the leaves during May and June, and the larvæ, on hatching, make their way to the stem, congregating and feeding beneath the leaf-sheath just above the first or second node. The point of attack is always low down on the plant, moderately close to ground-level. The larvæ, when fully grown, are about one-eighth of an inch in length and yellowish-white in colour, with no definite head. Pupation takes place in the position where the larvæ have been feeding, and a case similar in appearance to a puparium is formed, rather broad and flat in shape and of a chestnut-brown colour. From its similarity in appearance to the seed of flax (*Linum*) it is spoken of as the ' flax seed ' stage. Whether or not a second generation of flies appears in the year is not certain. If this is the case the flies develop again to the pupal stage before winter, which is passed in this way, the adults appearing again in the following May.

Plants attacked and damage done.—Grasses are the normal host plants of this insect, and the 'flax seed' stage may be found on Cocksfoot and probably on other grasses. Among cereals wheat, barley and rye seem to be preferred. The feeding of the larvæ causes a weakening of the straw at the point of attack, the result being that not only is the

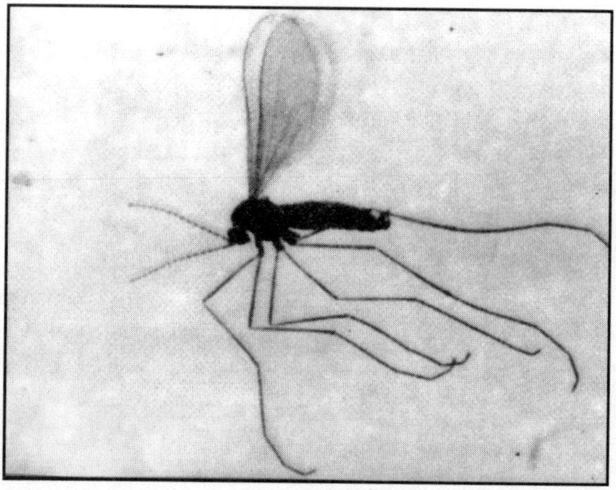

FIG. 109.—Hessian Fly (*Cecidomyia destructor*).

grain improperly filled but the crop very readily 'goes down,' the straw being elbowed over where it is weakened.

Natural Enemies.—The larvæ are subject to considerable parasitism by Hymenoptera, and probably the numbers are checked in this way so effectively that serious attacks are rare. For some reason during 1886 and 1887 a great deal of damage was caused in England and Scotland, but since then little has been heard of the insect as a pest.

Control Measures.—In consequence of its spasmodic appearance little has been done in this country in testing out means of controlling the Hessian fly. The majority of the pupæ remain in the stubble after the crop has been harvested, since this stage is usually found very low down on the plant. Where practicable, burning of a severely infested stubble would be effective, though of course this is not only a difficult operation to carry out, but is impossible if 'seeds' have been sown. Possibly the early sowing of spring cereals might help to overcome this attack, as in the case of the Frit fly (p. 217), where the fly seems deliberately to choose the smaller and less well developed plants for oviposition. Notably in America, where this pest is a serious one, the country has been mapped out in areas according to the periods when the adult flies are present, so that the 'fly-free periods,' as they are called, are known. The farmer is advised not to sow until after this, by which time the adult flies will have died and oviposition will not take place in the crop.

Grain from a badly infested crop should not be used for seed, since its poor development may result in bad germination and the consequent production of weak seedlings.

The Wheat Midges (*Contarinia tritici* and *Sitodiplosis mosellana*.

Description and Life History.—There has formerly been much confusion as to the species of midges found in wheat ears, and *Contarinia tritici* has usually been the one recorded. It has now been established, however, that it is no more common, as a rule, than a similar midge, *Sitodiplosis mosellana.* The two are quite easy to distinguish in all stages by their colour. The eggs, larvæ and adults of *Contarinia* are golden yellow in colour, while those of *Sitodiplosis* are a carroty red.

Both species are approximately the same size, 2 mm., and occur at about the same time. They may often be seen hovering in clouds over a wheat field when the plants are flowering, and all the individuals appear to be females. Males are very rarely found in the field.

The eggs are laid in the developing florets and the larvæ feed there, sometimes preventing the formation of grain entirely, or causing it to be shrivelled and badly filled.

Natural Enemies.—Many parasitic Hymenoptera are associated with these midges and the numbers are checked in this manner. A great number of larvæ are sometimes found in the grain at harvest time, and these appear all to be parasitised, the healthy larvæ normally leaving the grain before harvest and falling to the soil, in which they pupate.

Control Measures.—Little can be done to prevent the damage by wheat midges. The advice formerly given, to burn the chaff, containing larvæ, which comes from the threshing-box, seems to be erroneous, because these are the larvæ which are parasitised, and such destruction would lessen the number of parasites. Probably cultivation of the wheat stubbles immediately after harvest, exposing the larvæ to the influence of weather and the attacks of birds, provides the most effective way of keeping down the attack.

The Pear Midge (*Contarinia pyrivora*).

Description and Life History.—This fly is blackish in colour and about one-tenth of an inch long ; the female has a long ovipositor. The eggs, which are long and white, are laid in batches in the pear blossom, often before it is open. This takes place during April and early May and the larvæ at once proceed to feed, causing at first a rapid and somewhat swollen growth of the fruitlet. As growth proceeds, however, the pears become distorted, and the majority of those attacked fall off the tree while still very small. Throughout June, attacked fruitlets may usually be found beneath the tree. Inside these may be found a decomposing black mass of frass and often numerous yellowish larvæ (Fig. 110). Many larvæ leave the pears before they drop, escaping by way of cracks or rotting areas, which usually are present. Those which are unable to escape remain in the pears until they rot, but the larvæ, once on the ground, work their way beneath the surface and pupate in a tiny, whitish cocoon, from which the flies emerge in the following April.

Plants attacked.—Only pears, so far as is known, are attacked, and of these the earlier varieties, such as Williams' Bon Chretien, are more liable to attack than others. The crop is often seriously depleted and may be entirely lost. The falling of young fruit is often attributed to many other causes, but there is reason to believe that this insect is more widespread and responsible for much more damage than has been supposed.

Control Measures.—Where possible all attacked fruitlets should be collected and burned, but this is only practicable on a small number of trees and could not be readily carried

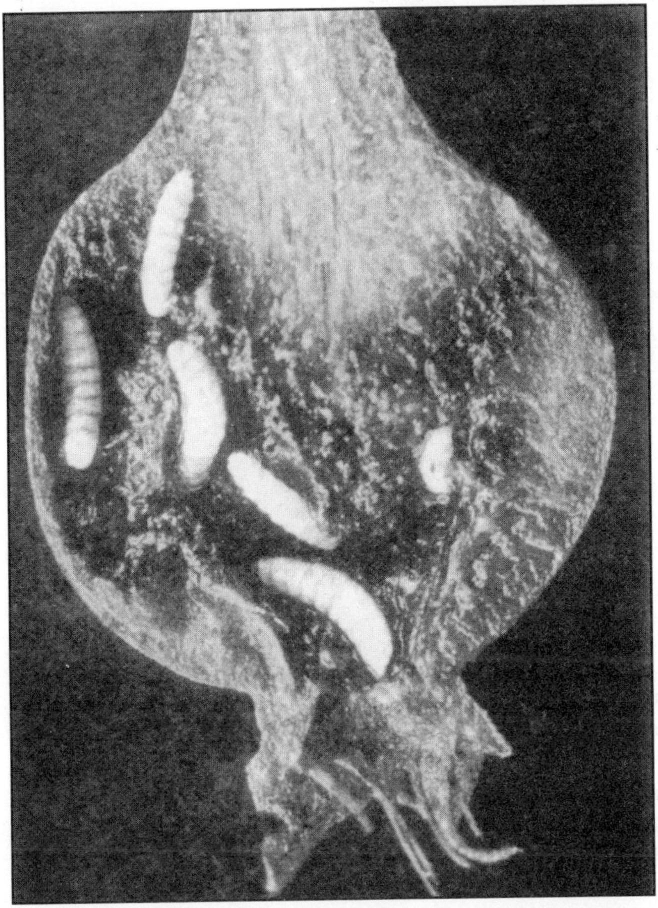

FIG. 110.—Larvæ of the Pear Midge (*Contarinia pyrivora*) inside a pear.

out in an orchard. Cultivation beneath the trees helps to expose the larvæ and pupæ, but this of course cannot be done if there is permanent grass. Kainit has been successfully used by spreading it evenly on the ground beneath the tree when the larvæ are making their way out of the pears and burying themselves in the soil. Dressings of half a ton to the acre upwards have been applied in grass orchards, and, particularly in America, such dressings have been found to give good results.

SERIES 2.—*Brachycera.*—Little is known of many of the families in this series and few are of economic importance. In one family the adults are blood-suckers and prey upon warm-blooded animals, but others feed upon the nectar of

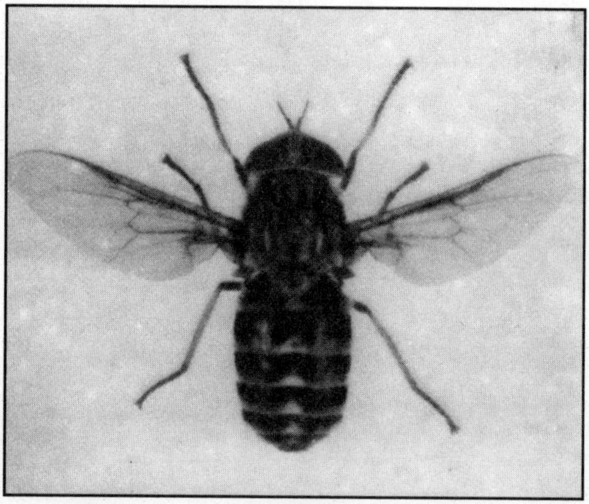

FIG. 111A.—Ox Gadfly (*Tabanus bovinus*). ♀

flowers. The larvæ live chiefly upon decaying animal or vegetable matter, though some are predaceous.

FAM. TABANIDÆ.

Gad Flies.

In this family we have flies which are of considerable economic importance because of their habit of sucking the blood of vertebrate animals. This habit is confined largely if not entirely to the females, the males living upon nectar. The larvæ are found in wet, swampy situations, either feeding upon rotting organic material or preying upon small organisms such as other insect larvæ. They have no definite head, and each body segment is supplied with a number of projections (pseudopods).

The blood-sucking habit of the adult females is well

known, both man and cattle suffering from these bites during the summer. There are three more or less common species in England, *Tabanus bovinus*, a large fly almost

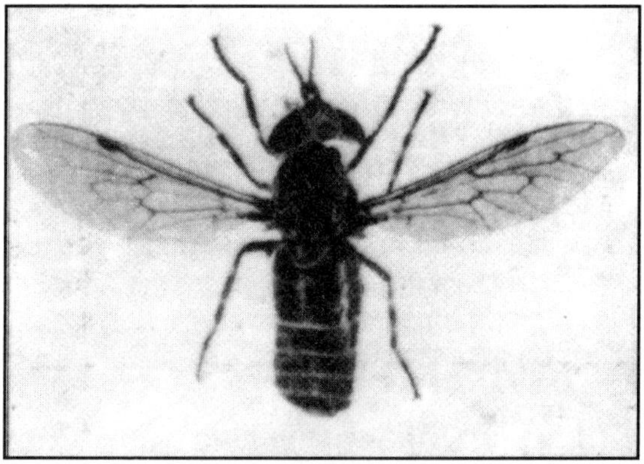

FIG. 111B.—" Clegg " or " Breeze " Fly (*Hæmatopota pluvialis*). ♀

FIG. 111C.—Blinding Storm Fly (*Chrysops cæcutiens*). ♀

an inch in length; *Hæmatopota pluvialis*, the well-known ' clegg ' or ' breeze fly,' our commonest Tabanid; and *Chrysops cæcutiens*, a green-eyed fly with dark markings on the wing, both the last mentioned being half an inch or rather more in length. All these Tabanids are somewhat flattened insects with large heads and well-developed piercing and sucking mouthparts. Some species produce a low audible hum when in flight, but others are remarkably silent, the first warning of their presence being the pain of the bite. Owing to their habit of causing stampeding among cattle and horses they are commonly known as Gad Flies. Stampeding may be the cause of serious loss of condition among young stock, and cows in milk may also suffer considerably. It is advisable to keep the cattle in during the hot midday, when these flies are most active, or to provide them with adequate shelter in the pastures. The flies are usually most troublesome during hot, thundery weather, and during the hottest hours around midday.

2. CYCLORRAPHA (SUB-ORDER)

This sub-order includes those flies which emerge from the pupa through a circular slit.

SERIES I.—*Athericera.*—This series contains those insects which we commonly think about when talking of flies, because the majority of them bear a superficial resemblance, at any rate in shape, to such common insects as the House-fly. The classification given is in many respects far from being satisfactory, but it is the one most commonly accepted at present, and it serves to sub-divide what is otherwise rather an unwieldy group.

SECTION A.—*Aschiza.*—The meaning of this term has been already explained (p. 196). Apart from one family the section does not contain any flies of much economic importance.

FAM. SYRPHIDÆ.

Hover Flies.

The insects contained in this family are moderately large flies, the majority of which must be regarded as beneficial, since their larvæ are predaceous and prey upon such pests as aphides. During the summer, when colonies of aphides are present upon plants and trees, one can usually find among them small, legless, flattened, yellowish or greenish-brown larvæ about a quarter of an inch in length. These readily devour aphides and are responsible for preventing their increase. They are the larvæ of Syrphid flies, and hatch from eggs laid by the adults near the aphis colonies. The winter is frequently passed in the larval state under cover of rough bark, etc., on trees, and pupation takes place early in the year. The family derives its name of the Hover flies from the habit of the adults in hovering in hawk-like fashion in the air, and then suddenly darting off a short distance and repeating the performance. Most of the species possess bright colouring, yellows and browns predominating, so that many of these flies look at first sight

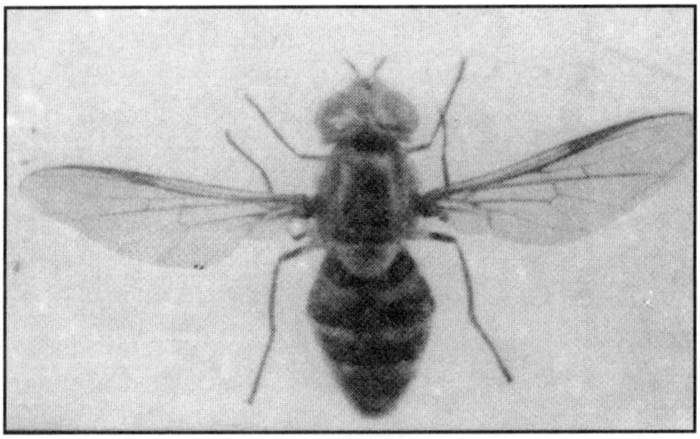

FIG. 112.—Hover Fly (*Syrphus ribesii*).

rather like bees and small wasps. They are commonly seen in flower heads, sucking nectar, frequently in the company of bees (Fig. 112).

Hover flies (Syrphidæ), Blood-sucking flies (Tabanidæ, etc.) and Blow flies (Muscidæ) are very similar in size, and are frequently confused. They may be distinguished by the venation of the wings, which is shown in Fig. 113. In blood-sucking flies all the veins extend to the margin of the wing (*A*); in hover flies there is a vein running parallel to the apical or outer margin; there is also a *vena spuria* or vein-like thickening running diagonally across the wing, which is present in no other group of flies. It is less distinct than true veins, and is not associated with them (*B*); in blow flies the apical region shows decidedly zigzag nervures (*C*).

Syrphid flies are sometimes accused of biting cattle, but this they are unable to do owing to the structure of their mouthparts: they are pollen and nectar feeders.

Two species, however, must be regarded as pests, and these are frequently called the Narcissus flies.

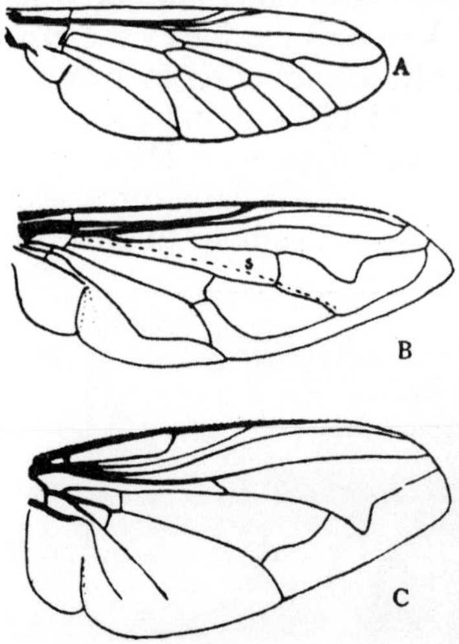

FIG. 113.—*A*. Wing of Hæmatopota. *B*. of Eristalis. *C*. of Calliphora. *S*. Vena spuria.

The Large Narcissus Fly (*Merodon equestris*) (Fig. 114).

Description and Life History.—In size and general appearance this fly is very like a small bumble bee, but of course may readily be distinguished by the fact that it has only two wings. It is very hairy, with brownish and yellow bands of colour on the abdomen, which add to its bee-like appearance. The adults are on the wing during spring and early summer, and eggs are laid on the neck of the bulb of the narcissus (and other plants). The larva on hatching bores its way into the bulb, but the way in which it effects this entrance is not certain. It may either do so by working in between the bud scales, or by working underground and entering at the basal ring, from which the roots arise. In attacked bulbs, at all events, the basal ring is usually damaged. Inside the bulb the larva feeds and grows into a stout, whitish grub about half an inch in length, with no definite head and no legs. It is usually found at this stage near the neck of the bulb. Pupation may take place in the autumn, but in the majority of cases the larvæ remain in the bulbs until spring, when they pupate either among the rotting 'remains' or in the adjacent soil.

Plants attacked and damage done.—This insect is abundant on the Continent and in most bulb-growing areas in England. Imported bulbs are sometimes found to be affected, and the pest has largely been spread in this way. Narcissus, hyacinth, amaryllis and many other bulbs are liable to attack. Part of the destruction is caused by the feeding of the larva itself and part by other organisms which thrive on damaged bulbs. The larva is, however, the primary cause, and the damage commenced by it is speedily turned to complete destruction by other organisms which gain an entrance. Attacked bulbs often fail to grow, and gaps occurring in a row should always be investigated and the cause ascertained

Dried bulbs containing large larvæ are usually found to be soft around the neck. It is almost impossible to detect the presence of small larvæ from an external examination of the bulb, though any sign of damage to the basal ring is suspicious.

FIG. 114.—Large Narcissus Bulb Fly (*Merodon equestris*). ♀

Control Measures.—Attention should always be given to bulbs which fail to grow, since these probably contain larvæ. If this is found to be the case the bulbs should be burned. It is useless to throw out such bulbs, since the flies will hatch from them.

If the bulbs are steeped in tepid water for twenty-four to forty-eight hours, it is found that most of the larvæ are destroyed, being either squeezed out when the bulb swells, or drowned. The latter is not likely to be the case, however, since the larvæ are capable of existing for a long period under water. It is important not to allow the water in which the bulbs are immersed to become very cold or they may be damaged. No very satisfactory treatment, however, has yet been discovered, though hot water sterilization (see p. 291, Bulb Eelworm) should be successful. Owing to the fact that bulbs are so widely grown in gardens, and that many are left in for several years before being dug up and separated, it is doubtful if damage can be prevented except by means which will prevent oviposition on the bulbs.

The Small Narcissus Fly (*Eumerus strigatus*).

Description and Life History.—This fly is about a quarter to one-third of an inch in length, almost black in colour, with a shiny metallic-green reflection: there are narrow white markings on the abdomen. Throughout May and June the fly is on the wing, and eggs are probably laid on the neck of the bulbs among the dying leaves, often a dozen or more to each bulb. The larvæ gain an entrance through the neck, and a very rapid rot follows; the whole bulb becomes a decayed mass in a few weeks. The larvæ are much smaller than those of *Merodon*, legless, and whitish in colour, with a prominent hard chestnut-coloured tubercle at the tail end. Large numbers may be found in one bulb, often with a larva of *Merodon*. In the autumn some of the larvæ pupate in the decayed mass, but others do not pupate until the spring.

FIG. 115.—Small Narcissus Bulb Fly (*Eumerus strigatus*). ♀

Plants attacked and damage done.—Narcissus and hyacinth are most commonly attacked in this country, but on the Continent it is reported that the larvæ are sometimes found in onions.

Control Measures.—The complete life history of this fly is still in doubt, and its habits are not well known. Apart from the rigorous destruction of bulbs containing larvæ, nothing is known which will prevent the attack.

SECTION B.—*Schizophora.*—As indicated on p. 196, it is usual to divide this section into two super-families, the *Acalypteræ* and *Calypteræ*.

SUPER FAMILY.—I. *Acalypteræ*

FAM. OSCINIDÆ (*Chloropidæ*).—This family is represented by a number of species which are common among herbage, and they are mostly small flies whose bodies are devoid of hairs. Two of the flies, the Frit fly and the Gout fly, are well-known pests of agricultural crops.

The Frit Fly (*Oscinis frit*).

Description and Life History.—The fly is about one-eighth of an inch in length, almost black in colour, and the female is slightly larger than the male. The wings from tip to tip measure about three-sixteenths of an inch, so that in the field the fly is not very conspicuous (Fig. 105, *B*). The adults appear about the middle of May, and elongate, white eggs, just visible to the naked eye, are laid low down on grasses and cereals. In a few days the eggs hatch, the larvæ at once boring into the shoot, where they feed in the region of the growing point, leaving no obvious sign of damage to be seen from the outside. The larva is white in colour when newly hatched, becoming pale yellow as it grows older. It may be found head end upward inside the shoot, which soon begins to show unhealthy symptoms. The tip of the central shoot begins to wilt and gradually dies back, eventually becoming yellow and dry, while the outer leaves remain quite green. If the plant be in a rather less advanced stage of growth, before the central shoot has appeared, then the vertical growth of the plant is checked, the region around ground-level becomes swollen, and secondary tillers are formed, a low, spreading growth resulting. No central shoot can be seen at all in this case, since it is killed out before it appears. In a few weeks the larvæ are fully grown, and pupate either on the outside of the plant or just inside the surrounding sheath, but occasionally on the soil in the immediate neighbourhood. Typical brown puparia, about one-eighth of an inch in length, are formed, and from these a second generation of flies hatches out. By this time the panicles are about to break free from the leaf-sheath, and eggs are laid by this second generation of flies inside the glumes, the larvæ boring their way into the developing grain, where they feed on the starchy material. Puparia may be found in the grain just about harvest time, but many fall to the ground as the panicle unfolds. A third generation hatches during August or September, and these flies usually lay their eggs on wild grasses, choosing young growth for the purpose. A great number of eggs may be laid on the rye grass in grass and clover leys. This fact is very important in view of another kind of attack which is sometimes severe. The larvæ, on hatching from these eggs laid in August and September, feed on the grass shoots, which show characteristic signs of damage. They remain as larvæ throughout the winter, pupating in the following March.

If these larvæ are very prevalent in a rye grass ley which is ploughed up for wheat, then the larvæ may migrate from the ploughed-in grass into the growing wheat. During January and February signs of this attack may often be seen, a serious thinning of the plant resulting. Ploughing-in appears not to kill the larvæ, which feed on the grass while it remains undecayed. It has been found that flies can emerge from buried pupæ through several inches of tight soil.

Plants attacked and damage done.—A great number of wild grasses are utilized by this fly, among which are the rye grasses, upon which the complete annual cycle may apparently be carried out. Among cereals, spring oats usually suffer most damage, but occasionally barley and rye are attacked (Fig. 116). The larva, feeding on the central shoot, destroys a potential panicle, and though side tillers may spring up they never produce as much grain as the original ones, and the crop ripens unevenly. In the first place, then, the attack results in a marked reduction of the number of panicles produced, and consequently on the weight of crop. Wheat, which is usually autumn sown, is not attacked except by larvæ which have migrated from ploughed-in grass, and winter oats are seldom damaged, a point which should be noted, in view of possible control measures.

Cereals, particularly rye, are sometimes sown in August or September for early feeding purposes, and the flies which emerge then may oviposit on such plants, while 'volunteer' plants in stubbles are sometimes used in the same way, taking the place of wild grasses.

Natural Enemies.—The degree to which this insect is subject to parasitism is not definitely known, but some causes seem to operate which at times check its increase in numbers. A certain amount of damage may always be seen in spring oats, but in some seasons it is much more severe than others. Climatic conditions may have some effect on this, a point which will be mentioned again. Some hymenopterous parasites can be bred from the puparia of the Frit fly, but they are usually not numerous.

FIG. 116.—Oats damaged by Frit Fly.
Note the reduction of the tillers at the base of the main culm.

Control Measures.

Spring Oats.—It has been observed that winter oats seldom have eggs laid on them, and consequently do not suffer. The probable reason for this is that the flies select young or backward plants for oviposition because the larva is better able to gain an entrance. The fact that earlier sown spring oats are less attacked than those sown later tends to strengthen this view, and experiments carried out on the University Farm at Cambridge in 1923 show that in early sowing lies the key to control.

Plot No.	Rate of Sowing.	Date Sown.	Yield in Cwts. per Acre. Grain.	Straw.
1	3 bushels per acre	Feb. 1st	16·0	42·7
2		Feb. 20th	14·9	31·0
3		Mar. 1st	15·2	29·3
4		Mar. 15th	10·7	25·1
5	4 bushels per acre	Mar. 23rd	9·9	25·3
6		Mar. 31st	6·2	18·1
7		Apr. 14th	2·2	8·8

The above plots were on a light gravel soil and all received exactly the same treatment except that they were sown on different dates. Such early sowing is not possible in all seasons and on all soils, but every effort should be made so that the plants are well established and growing vigorously by early May when the eggs are laid. The small yield of grain in the later sown plots was due to the fact that the growing points were killed out and less panicles were consequently produced. If the number of ears on Plot 1 above be taken as 100 per cent., the numbers on Plots 6 and 7 were 30 per cent. and 10 per cent. respectively. The diminution in the number of ears was directly proportional to the increase in the number of tillers attacked by the larvæ.

The sample of grain from early sown oats is also better than that from later sown ones, since the attack on the grain by the second generation of flies is less. A bad attack of this nature has been found in extreme cases to reduce the bushel weight of oats to about 24 lbs. instead of the normal 40 lbs.

Winter Wheat after a ploughed-up ley.—In preventing this attack it is essential to plough the ley as early as possible so as to allow a long period to elapse before the wheat is drilled. Ploughing should be done by harvest if possible, or immediately after, and wheat sowing left as late as practicable.

The Gout Fly or Ribbon-footed Corn Fly (*Chlorops tæniopus*).

Description and Life History.—In size this fly is about one-eighth of an inch long with about a quarter of an inch expanse of wings (Fig. 117). It has a yellowish head and thorax, the latter being marked with three longitudinal black markings on the dorsal surface. The abdomen is greenish-black with darker transverse markings. Two generations occur yearly in this country, a summer one which lays eggs on cereals, usually barley, and an autumn one which oviposits chiefly on 'twitch' or couch grasses. The adult flies normally appear toward the end of May and early in June. Eggs are laid on the leaves of barley, and the larvæ on hatching work their way downwards in the leaf-sheath until they come to the developing ear. They travel beyond this as far down as the first node below the ear and then turn upward again, by which time they are fully grown. Pupation takes place inside the leaf-sheath in the neighbourhood of the damaged ear, a typical brown puparium being formed. From this the fly emerges during August and eggs are again laid, this time on couch grasses. The larva lives in this shoot throughout the winter, pupating in the following March. From these pupæ the flies of the summer generation arise.

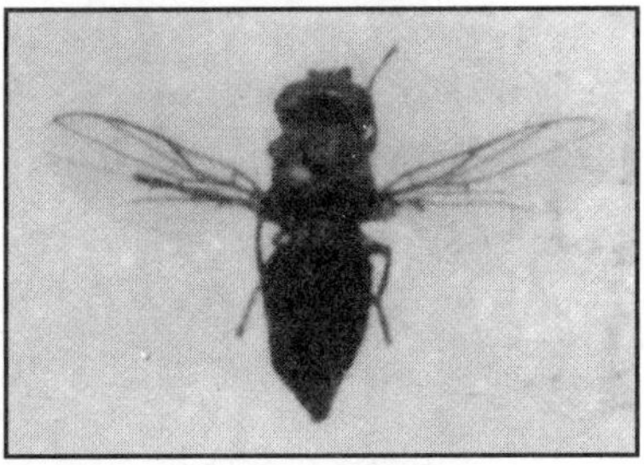

FIG. 117.—Gout Fly (*Chlorops tæniopus*).

Plants attacked and damage done.—Couch grass is the normal winter host of the fly and is also attacked at times during the summer. Winter barley may also be attacked by flies of the autumn generation. Barley is the spring-sown cereal most commonly attacked, though damage has been recorded to wheat and rye. Shoots of couch grass which are attacked are very much shorter than normal, very swollen, and of a deep green colour. The same appearance is characteristic of the attack on winter barley. In the case of spring barley the plant, having grown considerably by the end of May, has an erect habit, and the ear-bearing shoots are several inches high. It is on the leaves arising from these shoots that the eggs are normally laid, the larva invariably working downwards when it gets inside the leaf-sheath. It commences to feed on the ear, which, if very young, is entirely killed out, but if in a rather more advanced stage of growth, has one side destroyed, the other being partially developed. Having passed the ear, the larva cuts a deep groove in the internode below, thus cutting off the sap supply to the parts above. The growth of the shoot is arrested, the leaves take on a twisted or distorted shape, and the ear never

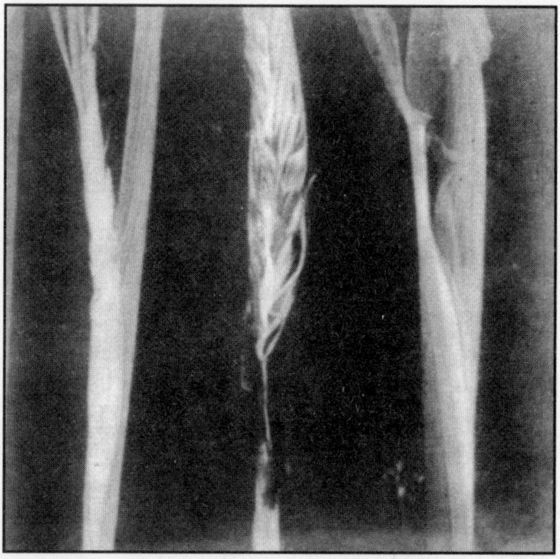

FIG. 118.—Barley Ears damaged by Gout Fly.

emerges from the leaf-sheath, which assumes a characteristic swollen appearance. It is this which gives the name of Gout fly to the insect. The larva by this time is almost fully fed and the damage has been done. Attacked ears seldom produce any well-developed grain, and must be regarded as a dead loss. They are usually so stunted that they stand far below the level of the normal ears, and the attack may be overlooked in the field, except for the fact that the crop looks 'thin.' Severe attacks, however, commonly result in 20 per cent. of the ears being totally destroyed (Fig. 118).

Natural Enemies.—The year 1887 is recorded as being

particularly bad in respect of this fly, and at intervals since then severe attacks have occurred. Some small amount of damage may usually be found every year, but some factors must be at work which normally keep down the numbers of this fly. Probably hymenopterous parasites play a large part, for a considerable proportion of larvæ are often found to be parasitised, and at least four different species parasitic on this fly are known.

Control Measures.—The number of flies in any given area is to some extent dependent on the presence of suitable winter hosts. Couch grass is a universal weed, if not in arable land, in hedge bottoms and waste places, and its complete eradication, however desirable, is impossible. It must be assumed, therefore, that there will be a summer generation of flies to lay eggs upon barley, whatever measures are taken. It has been suggested that the Frit fly instinctively selects suitable shoots on which to lay its eggs, and there is even more evidence that this is the case with the Gout fly. The larva works downwards in the leaf-sheath, coming from an egg laid on the leaf. It has been already pointed out that the ear, and the stem immediately below, is the locality in which it feeds. If, therefore, the plant is so forward in growth that the developing ear inside the sheath is at least as high up as the top leaf when the eggs are laid, then the possibility of a larva reaching the ear is very small.

The diagram on the next page will serve to make this clearer (Fig. 119).

A and *B* represents two barley plants in different stages of growth. *E* is the region of the developing ear relative to its position in the sheath. Leaves are represented at 1, 2, 3, 4 and 5.

In *A* it will be seen that eggs laid on leaves 1, 2 and 3 will give rise to larvæ, which, travelling downward, will reach and destroy the ear, *E*. Such leaves have been called 'critical leaves,' as distinct from 4 and 5, which are non-critical. In case *B* there are no critical leaves, and this ear will escape damage. Eggs as a rule are laid only on critical leaves, very few being found on non-critical ones.

The whole problem of control, therefore, depends on the

stage at which the plant has arrived by early June, when oviposition takes place.

Any factors which tend to retard growth predispose the crop to attack, and any which accelerate it tend to give immunity. Cultural conditions are therefore all-important. A fine, firm seed-bed should be aimed at with the land in as good 'heart' as possible, to ensure continuous unchecked growth. Suitable manuring may assist this, though care must be taken not to force the crop so much as to spoil the sample of grain or to make it 'go down.' Superphosphate is a fertiliser which assists in promoting early maturity, and it has been found very beneficial in this way.

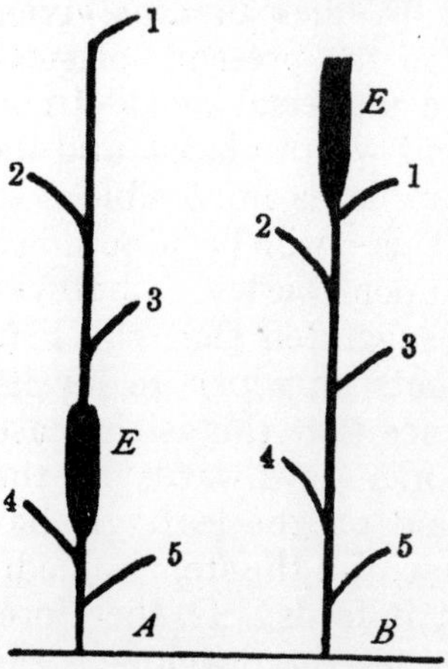

FIG. 119.—Diagram showing critical and non-critical leaves of barley. For explanation see text.

FAM. TRYPANEIDÆ (TRYPETIDÆ).

The Celery Fly (*Acidia heraclei*).

Description and Life History.—The prevailing colour of this fly is light brown, with the exception of the eyes, which are green, and there are darker brown bands on the wings. The fly is about one-quarter of an inch long, with a wing expanse of three-eighths of an inch (Fig. 120). The eggs are white in colour, the larva yellow or greenish, and the puparium of a similar colour.

Eggs are laid on the leaves, usually on the under side, and the larvæ on hatching immediately bore in, feeding on the chlorophyll tissue between the upper and lower epidermis. When fully fed they fall from the leaf and bury themselves in the soil, where they pupate. Two or more generations may occur in the year. The fly first appears during early May, and the last generation of the year over-winters in the soil in the pupal state.

Plants attacked and damage done.—Celery is the chief cultivated plant which suffers, but occasionally the leaves of parsnips are attacked. The larvæ are also to be found in many wild Umbelliferæ. The feeding habits of the larvæ result in a reduction of the leaf area of the plant. Attacked leaves show large blistered patches where the chlorophyll tissue has been destroyed, and the growth of the plant is consequently checked. The worst results follow an attack on the seedlings, when severe stunting or death may result. After the plants have been put out, successive attacks may occur, all checking the normal growth.

Control Measures.—Little work has been done on this fly which has yielded satisfactory results. Seedlings in boxes may be covered with muslin or other fine material to preserve them from attack. Once the plants have been put out they cannot be protected. They grow vigorously if given plenty of water or liquid manure, but this only compensates for the damage done by the fly, in forcing new leaf growth. The larvæ may be pinched in the affected leaves as soon as the attack is noticed, but this is not practicable on a large scale. Deterrents such as soot and lime scattered on the leaves from below may be effective for a short time, but no certain method of control has yet been found.

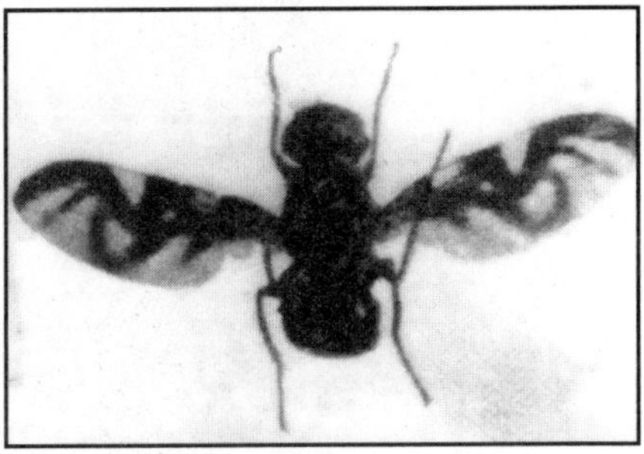

FIG. 120.—Celery Fly (*Acidia heraclei*). ♀

FAM. PSILDÆ.

The Carrot Fly (*Psila rosæ*).

Description and Life History.—In length the Carrot fly measures about one-quarter of an inch and the wing expanse about three-eighths of an inch. The body is of a shiny greenish-black colour, and the tip of the abdomen is pointed in the female. Iridescent colours may be seen on the wings, but there are no definite bands of colour on them. The larvæ when fully grown are rather more than three-quarters of an inch in length and yellowish in colour. The puparia are brown.

Adult flies are on the wing during late May, and lay eggs

around seedling plants just below soil level. The larvæ work their way down beside the carrot, frequently cutting through the tip of the root. Others bore galleries just into the flesh of the root, causing rusty coloured tracks to appear. Attacked roots when pulled up frequently have larvæ projecting from these burrows, the head end being

FIG. 121.—Carrots damaged by Carrot Fly (*Psila rosæ*).

embedded in the carrot. When fully grown the larvæ pupate in the adjacent soil and two or more generations may occur in a year. Puparia of the last generation remain in the soil throughout the winter.

Plants attacked and damage done.—Carrots and celery among cultivated plants are attacked. If, as frequently happens in the case of small plants, the tap root is cut through, the plant dies and whole beds may be killed off in this way. Those plants which develop further are stunted and disfigured by numerous galleries bored in them.

Attacked carrots do not keep well in storage and produce badly shaped roots, a great deal of which has to be cut away. Attacks by this fly are usually much more serious where carrots are grown on a small scale than in fields (Fig. 121).

Control Measures.—Very shortly after the plants are thinned, signs of attack frequently appear. The process of thinning results in the loosening of soil around the plants, and the fly is able to lay its eggs in suitable positions. It is, therefore, advisable to sow thinly and to thin out where necessary as early as possible, consolidating the ground around the remaining plants when this has been done. Light land which has been deeply trenched and has not received any farmyard manure since the previous autumn is most suitable for carrot cultivation, and crops on such land are less liable to severe attacks.

Deterrent substances, such as paraffin mixed with soot, ashes or sand, scattered along the rows, are found to be effective in some cases, and dressings of salt are also said to be beneficial.

SUPER FAMILY.—2. *Calypteræ*

FAM. ANTHOMYIDÆ.—The flies in this family are moderately large in size, somewhat resembling a common House fly. They are usually grey or black in colour. The larvæ of many species are carnivorous; others are to be found in decaying organic matter, while some feed on living plant tissue.

The Wheat Bulb Fly (*Leptohylemyia coarctata*).

Description and Life History.—The length of the body in this fly is about three-eighths of an inch and the wing expanse rather more than half an inch. The prevailing colour is dark grey and the fly is covered with fine hairs.

Adults are on the wing during the summer months, emerging early in July, and the eggs are deposited in bare soil, particularly if it is dry and dusty. This singular method of oviposition was discovered when it was found that in almost every case attacks occurred on wheat sown after a bare fallow, or on land from which a crop of early potatoes had been lifted. The eggs, which are laid in July and August, remain unhatched until early in the following year. The larva bores into the young wheat and may be found during March feeding inside the shoot in the region of the growing point. Its position is similar to that of the Frit fly larva, and the appearance of the attacked plant much the same also. When fully grown the larva is about three-eighths of an inch long and whitish in colour. It pupates in the soil at the depth of about two inches, a typical brown puparium being formed. Only one generation occurs in the year.

Plants attacked and damage done.—Wheat is the main crop which suffers damage, but rye and winter barley are sometimes attacked. It appears that almost any cereal crop is liable to attack if sown on land which has been bare, entirely or partially, during the preceding August. Thus wheat may be attacked following on a bare fallow, potatoes or roots, particularly if they have been poor crops and not covering the ground well.

The effect of the larval feeding is to destroy the growing point of the plant. Less ears are produced and the weight of crop is consequently diminished.

Control Measures.—The obvious method of preventing these attacks is to have the ground covered by some crop during August if possible. A fallow, if clean by then, might be sown with mustard or other crop suitable for turning in before wheat-sowing. Failing that, it is unwise to have a fallow very fine, for eggs are not laid to the same extent on rough, cloddy soil as on a fine tilth. A half fallow overcomes the difficulty to some extent. An alternative method is to alter the rotation so as to take wheat in some other place than after a fallow, though this, of course, has its disadvantages; the practice should be adopted which best fits in with the type of farming.

The Onion Fly (*Phorbia cepetorum*).

Description and Life History.—In size and colour this fly is very like the preceding one. The prevailing colour is grey, but there are darker markings on the thorax and abdomen (Fig. 122).

Eggs are laid during May on the neck of the onion plant or in the adjacent soil. Several eggs may be deposited on each plant, and these hatch in about seven days. The larvæ bore their way into the bulb and are fully grown in about three weeks, when they pupate in the decaying bulb or in the soil. Adult flies emerge in two or three weeks from these pupæ and eggs are again laid. Probably three such generations occur in the year, the last remaining in the pupal stage over winter. Larvæ may sometimes be found in bulbs very late in the year, and in America it is said that the adult hibernates.

Plants attacked and damage done.—Onions and shallots

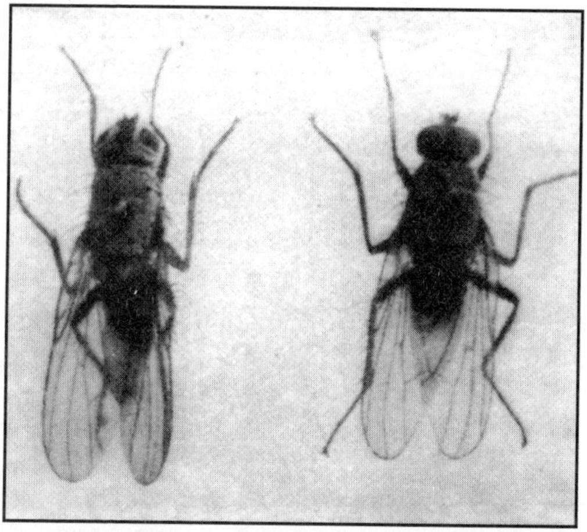

FIG. 122.—Male (left) and female Onion Flies (*Phorbia cepetorum*).

are usually attacked, and spring-sown onions suffer more than those sown in autumn. Attacked plants show yellowing of the tips of the leaves, which may readily be pulled away, having rotted just at ground level. The leaves may drop over and sometimes they become swollen and flabby. Such plants fail to produce good onions. The bulbs rot away and those bulbs attacked at a later stage rot in storage.

Control Measures.—It has been noted that autumn-sown onions are less attacked than spring-sown. The latter should be pushed forward in growth by good cultivation methods and manuring. Preventive dressings seem to be uncertain in their effect, but paraffin and carbolic acid emulsions have been used with good results. The adult female fly is known to feed on sweet liquids and in America efforts have been made to turn this fact to account. A syrup composed of molasses and water, poisoned with sodium arsenite, is exposed in vessels, but it is not definitely known how far this practice is successful. Onions grown in shallow trenches and earthed up appear to have escaped injury in some cases.

The Mangold Fly (*Pegomyia betæ*).

Description and Life History.—The prevailing colour of this fly is brownish grey and the legs are yellow. In size it is much the same as the onion fly (Fig. 123).

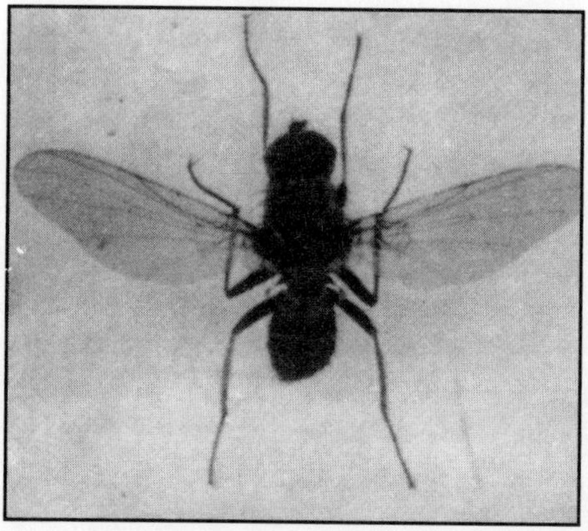

FIG. 123.—Mangold Fly (*Pegomyia betæ*). ♂

The adults appear during May, almost as soon as mangolds are showing through the soil. Batches of long, white, oval eggs, from three to seven or eight in number, may be seen attached to the under side of the leaves. In a few days these hatch, the larvæ boring into the leaves and feeding on the chlorophyll tissue in much the same way as celery fly larvæ. Whitish blistered patches appear where they have been feeding, and these areas finally turn brown and die, giving the foliage a scorched appearance. The larvæ are white in colour, and when fully grown are almost one-third of an inch in length. They fall to the soil and pupate, another generation of flies appearing in a few weeks. The larvæ do not all develop at the same rate, however, so that the generations are not very well marked. Probably three distinct ones occur, but all stages of the insect may generally be found in existence at the same time during the summer. The last pupæ formed remain in the soil and over winter there.

Plants attacked and damage done.—Mangolds, sugar beet and spinach among cultivated crops, and weeds of the same family, such as goosefoot, are attacked. The injury is due to the reduction in effective leaf area, and if this occurs in the seedling stage the plant may be killed. The damage is worst when an attack coincides with a spell of weather which checks leaf growth. Plants are not killed once they have passed the seedling stage but may be severely checked, with the result that the roots are small.

Control Measures.—It is found where farmyard manure is applied and ploughed under in autumn, that attacks are usually slight. The presence of fresh manure appears to attract this fly as well as many other Anthomyidæ. Any measures which encourage growth are useful, and the application of suitable top dressings sometimes helps the plants to grow away. Light rolling when the plants are in the seedling stage has been found beneficial in promoting growth and does no harm to the plants unless the land is very stony.

The Cabbage Root Maggot Fly (*Chortophila brassicæ*).

Description and Life History.—This is another fly very similar in size and shape to the Onion fly and differing only in detail.

Adults appear towards the end of April and early in May, and white elongate eggs are laid in the soil immediately around the roots of plants of the Cabbage family, both wild and cultivated. These eggs hatch in a few days, the larvæ at once boring down to the fibrous root system of the plant and feeding on the finer roots. These are eaten away and eventually nothing remains but the stump, which is frequently hollow and decayed. The plant dies unless it is able to make fresh roots as the others are eaten. When fully grown, the typical white maggot larvæ pupate in the soil. The puparia are reddish brown in colour. At least two generations occur per annum and the pupæ of the last one remain in the soil over winter.

Plants attacked and damage done.—A large number of wild Cruciferæ are attacked, including charlock and shepherd's purse. Among cultivated plants cabbage, cauliflower, brussels sprouts, turnip and radish frequently suffer. Plants on light, open soils appear to suffer most since they are most susceptible to drought. The root system is rapidly destroyed by the larvæ, first the fibrous roots being eaten off and finally the main roots. The leaves of the attacked plant take on a dull leaden colour and wilt in the sun, and unless new roots are rapidly pushed out the plant dies. In any case it is severely checked and may be almost worthless.

Natural Enemies.—Two small black 'rove' beetles (Staphylinidæ) often destroy large numbers of pupæ, and exercise some influence in keeping down the numbers of the fly.

Control Measures.—The critical period is in the early stages of growth of the plant, the attack by the first generation being the worst one. If the plant can make roots faster than the larvæ devour them, then it survives. Consequently any conditions assisting rapid root development are beneficial. The soil should be well manured, preferably some time before the plants are put out, and artificials given to stimulate growth. Firm soil is also essential.

As a deterrent, carbolic acid emulsion has been used, and in America good results have been obtained by the use of a .1 per cent. solution of mercuric chloride watered in. This solution acts as a larvicide but is extremely poisonous to man, and should be used with great care. A mechanical method has been devised which entails the use of a tarred felt disc, as shown in Fig. 124. This is three inches square and has two slits *AB* and *CD* cut in it.

When plants are put out, a disc is slipped round the stem of each one and pressed flat on the ground, care being taken to have the soil level underneath it so that the fly cannot crawl in. With such a disc in position, eggs cannot be laid close to the plant. Although the fitting of such discs entails a small amount of extra trouble, the results obtained fully justify their use, at any rate in gardens. Not only does the disc act mechanically in preventing oviposition but the shade it affords keeps the soil from drying out below and the plant roots more rapidly in consequence.

FAM. MUSCIDÆ.—The well-known House fly and the Blue Bottle fly are typical of this family. The majority of the species have mouthparts of a very specialized type, capable of taking in liquid food only, but the blood-sucking habit has been developed in some cases. Powerful piercing stylets capable of boring through the skin of animals are found in the Biting Stable fly (*Stomoxys calcitrans*), which otherwise closely resembles the House fly. No members of this family are plant pests, but the larvæ or adults may be animal parasites. Owing to their capacity for carrying the organisms which cause various diseases, many of these flies are important from the medical and veterinary point of view.

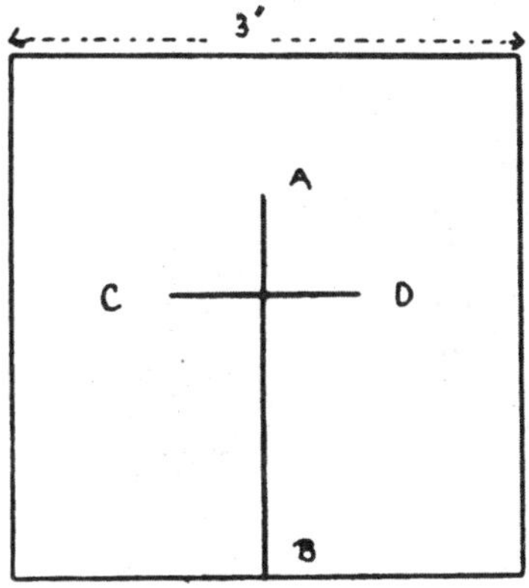

FIG. 124.—Diagram of Tarred Felt Discs used to protect Cabbages against the Root Fly.

The Common House Fly (*Musca domestica*).

Though it has no direct agricultural importance, the House fly is so common and serious a pest that it must receive mention. The eggs are laid in stable refuse or almost any decaying animal or vegetable matter, on which the larvæ feed. The adults find their way into dwellings and alight on food such as sugar, meat, or milk, contaminating it with bacteria carried on their legs. A number of diseases such as diarrhœa and typhoid are transmitted in this manner. Traps of many kinds have been invented for the purpose of destroying the adult flies. The treatment of refuse has also received attention, efforts being directed toward the destruction of larvæ in such breeding places.

The Sheep Maggot Fly or Green Bottle (*Lucilia sericata*).

Description and Life History.—This fly is of a brilliant metallic, green colour, about one-third of an inch in length, with a wing span of about three-quarters of an inch. A very similar fly is *Lucilia cæsar*, but this species seem to breed in carcases and seldom oviposits on live animals.

The Blue Bottle Fly (*Calliphora vomitoria*) is a closely allied fly, and also breeds in carrion.

The adults of the Sheep Maggot fly appear as soon as hot weather commences. Clusters of eggs are laid on the wool of sheep and these hatch rapidly, often in twenty-four hours. The larvæ work their way down the wool to the skin where they feed, causing a severe wound and great irritation. When fully grown they fall to the ground and pupate, a fresh generation of flies emerging in about a fortnight.

The eggs are whitish, about one-sixteenth of an inch in length. The larva is a typical white legless maggot, half an inch in length when fully grown, and it pupates in a brown puparium.

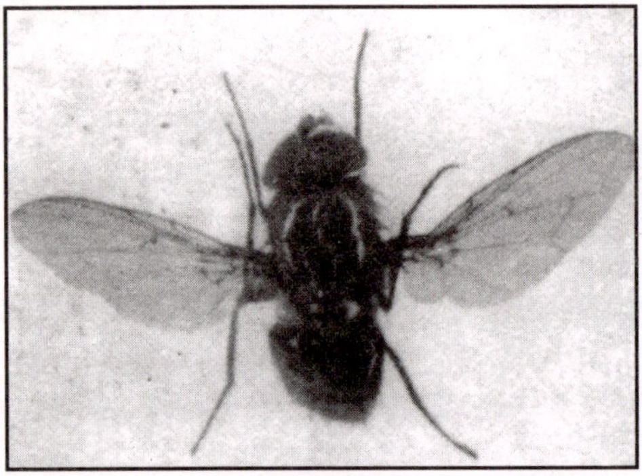

FIG. 125.—Sheep Maggot Fly (*Lucilia sericata*).

Nature of Damage.—Lambs are generally affected to a greater extent than old sheep. When 'struck,' a sheep shows evident signs of severe irritation, and an experienced shepherd is frequently able to detect such a sheep from a distance by its behaviour. Such animals thrive badly and are disfigured by loss of wool. In severe cases where it does not receive attention the sheep may die.

Control Measures.—Nothing is more important than cleanliness in preventing attack by these flies. Sheep which are scouring or otherwise have patches of dung-covered wool on the hind quarters, are most frequently 'struck,' and shearing round the tail and between the hind legs is beneficial. At times maggots are found on the back of the sheep, the wool matting together in a sticky mass due to fluids exuded from the wound. Care must be taken in treating animals with open sores of this nature, and damage may easily be done by the application of unsuitable dressings. By constant and careful attention, affected sheep may generally be picked out before severe wounds are caused. Sulphur ointments and powdered sheep dips rubbed well into the region where maggots are found will easily destroy them, but carbolic and arsenical substances should be avoided.

The effect of preventive dips or dressings does not last long, and expensive treatment of this kind is not generally worth while.

Equal parts of turpentine and rape oil form a good

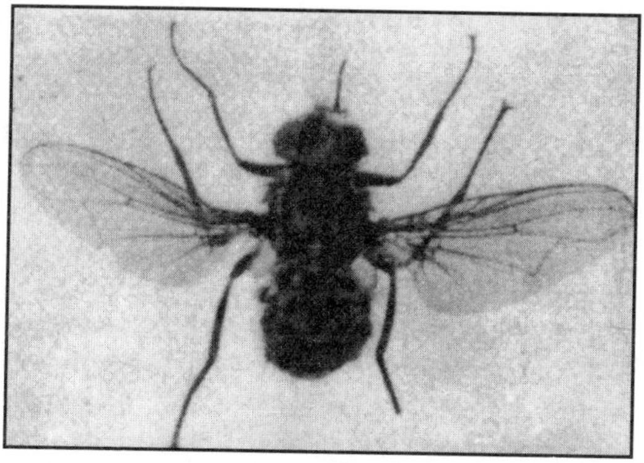

FIG. 126.—Biting Stable Fly (*Stomoxys calcitrans*). Note the sharp mouthparts.

mixture to apply where wounds are caused. As many maggots as possible should first be got out of the fleece, and after applying the oil, the affected part should be dusted over with flowers of sulphur.

The Biting Stable Fly (*Stomoxys calcitrans*).

It is commonly said that house flies bite when rain is imminent during the summer. Examination of the mouthparts of a house fly will show that it has no organs capable of biting (Fig. 101), and the insect responsible for such biting is the 'stable fly.' In size, shape and colour it much resembles the house fly, and breeds in similar situations. A careful look at the head end will at once distinguish these flies (Fig. 126). On the head of *Stomoxys* a horizontal forwardly projecting 'awl' is to be seen, and this encloses an elaborate set of piercing mouthparts. The blood-sucking habits of this fly are shared by other similar flies, such as the Tsetse fly of South Africa, but they are exceptional among the Muscidæ in possessing such mouthparts. Normally the Stable fly may be seen resting on sunny walls, but particularly in showery weather appears to take refuge in houses where it bites both man and animals. It is not regarded as a serious pest and no measures have ever been evolved for its suppression in this country.

FAM. ŒSTRIDÆ.

Warble Flies or Bot Flies.

This important family contains well-known flies whose larvæ are animal parasites. The adults are all large, hairy, bee-like flies characterized by having abortive mouthparts. The adults are incapable of blood-sucking although they are often confused with the Gad flies (Tabanidæ). Bot flies may cause stampeding among animals, but this probably is due to the low, buzzing sound they make while in flight, and not to the pain caused by bites.

Ox Warble Flies (*Hypoderma bovis* and *Hypoderma lineata*).

Description and Life History.—Although two distinct flies are responsible for 'warbles' they may be considered together, since, so far as is known, their life histories are very similar in all essentials. The adult flies are similar in size to the honey bee, nearly three-quarters of an inch in length, with a wing span of an inch. They are variously coloured, yellowish-brown and reddish-brown prevailing. *H. lineata* appears about the middle of May and *H. bovis* rather later. Both flies continue to trouble stock throughout hot weather until September. They are only on the wing during very warm days and are said to be so susceptible to sudden changes of temperature that they will not enter shade nor cross water. Unless careful watch is kept for them on suitable days they are seldom seen. The eggs, which are about one-twelfth of an inch in length, are laid on the hairs of the cattle, normally low down on the hind legs, and are attached firmly by means of claspers. The idea is prevalent that eggs are laid on the back since the warbles are found there, but it has been proved that this is not the case. In four or five days the little larvæ emerge, and working down the hair, penetrate the skin and the underlying tissues. Their movements from this point have not been definitely traced, but by some means they work toward the animal's fore end and have been found in the region of the œsophagus. They eventually leave this region, emigrating backward and upward in the muscular tissue until they appear in the back.

FIG. 127.—Ox Warble Fly (*Hypoderma lineata*).

Here they are found from January onwards. The usual situation is near the vertebral column, on either side. A swelling appears on the animal's back, perforated by a small hole which is partially concealed by a sticky clot of hairs. This hole is bored for breathing purposes by the larva, which is now almost three-quarters of an inch in length. It has a pair of spiracles at the tail end, no definite head, and its body is very wrinkled. During the early months of the year, when fully developed, the larva leaves the animal by the hole in the skin, falls to the ground and pupates under cover of a stone or clod of earth. The puparium is black in colour and the adult fly emerges from it through a cap which breaks off at the head end.

The pupal stage lasts from five to eight weeks.

Damage done.—The adult flies, although they do not suck blood, are responsible for injury to cattle. The milk yield from cows may be seriously reduced if they are 'stampeded' and similarly in-calf cows may suffer. Bullocks also lose condition under such circumstances. The larvæ, however, are responsible for the main damage. The flesh beneath the 'warbled' areas rapidly decomposes when the animal is killed and turns to a greenish, jellied mass unfit for food. The breathing holes made by larvæ in the skins reduce the tanning value, because they occur in the best part of the hide. Badly warbled hides may realize 3d. per lb. less than first-class samples, and in this manner alone it is estimated that about £500,000 worth of damage is done in a bad season.

Control Measures.—It is possible to deal with Warble fly damage in two ways. Any measures which would deter the fly from laying its eggs would be effective, but although many have been tried, no reasonable measure of success has been attained. It is difficult and expensive to apply substances to the legs of cattle, and, moreover, they are quickly rubbed off. The most hopeful results have been obtained by measures designed to kill the larvæ and so reduce the number of flies. To be really successful a method of this kind would need a considerable degree of co-operation between stock owners in any given area. The flies are moderately powerful on the wing, although they appear not to travel great distances, and re-infestation of an area from outside would gradually take place. Cattle imported from other districts may, of course, contain warbles, and the possibility of flies being introduced in this fashion must always be remembered.

Various dressings have been found to be effective in killing the larvæ *in situ*, and to some extent the method of squeezing them out is practised, but is not very satisfactory.

The most effective dressings are :

1. Tobacco Powder and Lime.—4 lbs. tobacco powder, 1 lb. lime, 1 gallon water.

The fresh lime is mixed in the water, the tobacco powder added and allowed to stand for twenty-four hours. The liquid is then strained through fine muslin and is ready for application.

2. Derris Powder.—This material is the powdered root of a tropical plant and has long been in use as a fish poison. It is mixed in the proportions of 1 oz. of powder to 1 pint of water.

3. Nicotine Sulphate and Lime.—Nicotine sulphate, 2 fluid ozs.; lime, 1 lb.; water, 1 gallon. The procedure in preparation is the same as in 1.

No harmful results follow the application of these dressings to the cattle, neither are the hides injured. The nicotine preparations (1 and 3) are best prepared freshly when required, as they do not keep well. Nicotine is poisonous and objectionable to some extent, so that there is much to recommend the second dressing, Derris, to the practical farmer.

The method of application of these dressings is important and one essential point must be emphasized. If the liquid is to be effective in killing the larva, it must touch it; consequently care must be taken to introduce it through its breathing hole in the skin. This is best done by means of a brass hand syringe with a fairly stout nozzle which is not liable to break. A fragile instrument such as a hypodermic syringe is unsuitable. It is desirable to clean off the mat of hair from the hole first by means of a damp cloth. The liquid may be applied also by squeezing it from a cloth or sponge or by means of a suitable brush.

On the Continent legislative action is now taken against the Warble fly by certain countries. Cattle owners must free their animals from warbles before a given date. The result of this is best shown in figures from Denmark giving the average percentage of warbled hides passing through the market:

1922—before legislative measures .	20½ per cent.
1923—1st year of the Act . .	–15 per cent.
1924—2nd year of the Act . .	–4½ per cent.

Suitable shade should be provided for cattle in pastures.

The flies will not readily enter shade and will not lay eggs on cattle there. It is a recognized practice among some stock owners to keep cattle in during the hottest hours around mid-day. Where the beasts have access to shallow water they will stand in it on such days and the legs are thus protected.

The Sheep Nasal Fly (*Œstrus ovis*).

Description and Life History.—The adult flies are yellowish-brown and hairy, having a striking superficial resemblance to a honey bee. They are about half an inch

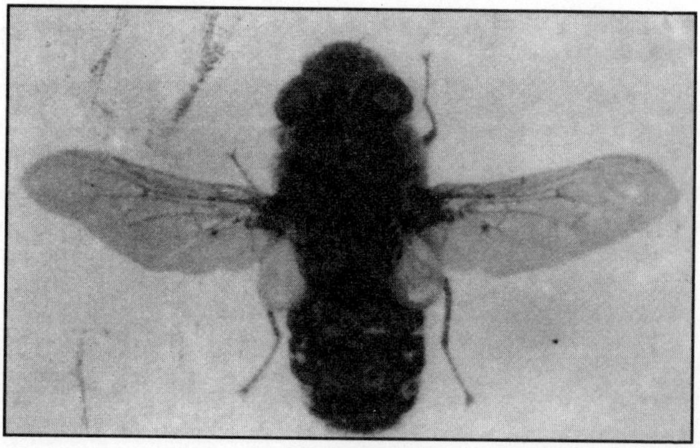

Fig. 128.—Sheep Nasal Fly (*Œstrus ovis*).

in length with a wing span of three-quarters of an inch, and are prevalent on hot days (Fig. 128). The eggs are white and somewhat kidney-shaped, and are deposited in the nasal passages of the sheep. Occasionally it seems that eggs are retained by the female fly until they hatch and then very tiny larvæ are deposited. In either case the tiny larvæ, by means of their mouth-hooks and numerous spines, work their way up to the nasal passages and may eventually get into cavities in the bones surrounding the fore end of the brain. When their development is completed the larvæ work their way downwards again and either fall or are sneezed out. They pupate beneath the shelter of clods or just beneath ground level, inside a puparium.

Damage done.—The irritation caused by the feeding of

the larvæ results in loss of condition of the sheep. The animals are very restless when the flies are about, often burying their noses in dusty patches, and frequently they stamp or paw the ground with their forefeet. In bad cases, when larvæ reach the region of the brain, the inflammation set up may cause pressure on that organ, resulting in a dazed condition. This is known as False Sturdy or False Gid, from its similarity to true gid caused by one stage of the tapeworm which lives in the brains of sheep.

Control Measures.—There are few practicable measures which can be put into operation. The larvæ cannot readily be killed except by fumigants, and one must resort to preventive measures. The noses of sheep are sometimes smeared with tar or fish oil, which are said to be repellent to the fly. Where salt troughs or boxes are provided for the sheep these may be so arranged that only a small aperture is left through which the animal can get its nose, and the edge of the hole is smeared with tar. This keeps fresh tar on the nose, since the sheep go constantly to lick salt. Whenever they are seen, the maggots should, of course, be killed. Only in rare instances is severe damage done, and only a few larvæ are usually present in one animal.

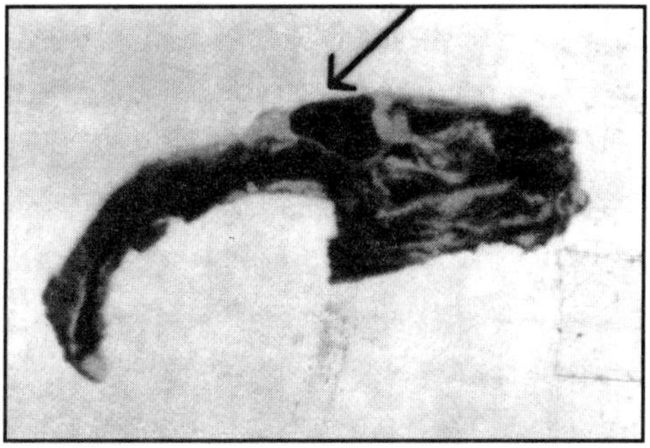

FIG. 129.—Larva of *O. ovis* in brain of sheep.

The Horse Bot Fly (*Gastrophilus equi*).

Description and Life History.—This fly is very similar to the preceding species in size and colour (Fig. 130). The tip of the abdomen is attenuated in the female, so that its actual length of body is greater than the others. Flies are on the wing during the summer months, most commonly in August. The eggs, which are yellowish in colour, are laid on the forelegs and shoulders of the horse and are attached to hairs. The animal licks off the eggs or newly-hatched larvæ, which thus pass into the alimentary canal, the young

FIG. 130.—Horse Bot Fly (*Gastrophilus equi*). ♀

FIG. 131.—Bots in lining of horse's stomach.

larvæ attaching themselves to the stomach wall by their anterior end. As many as thirty or more may be found together (Fig. 131). When fully grown they are almost three-quarters of an inch in length, rather barrel-shaped, and each segment is ringed by numerous spiny processes. Having completed their development the larvæ let go their hold, pass out of the horse with the fæces and pupate in the soil. The length of time occupied by the larval stage in the horse is about ten months, and the pupal stage lasts five to seven weeks.

Damage done.—No disturbance is caused by the presence of the flies, horses taking little or no notice of them. The feeding of the larvæ, if these are present in numbers, may cause serious loss of condition. The lining of the stomach is damaged and may become ulcerated, much the same as in the case of tapeworms.

FIG. 132.—Sheep Ked (*Melophagus ovinus*).

Control Measures.—As a rule no routine measures are taken against bots. Their presence is generally unsuspected until they are seen passing out of the horse, and any damage they have done is then past prevention.

SERIES 4. *Pupipara.* — All these flies are blood-sucking ectoparasites, and, with the exception of one species on bees, they have birds and mammals as their hosts. Some have wings, but these are not capable of sustaining long flights, and others are entirely wingless. The eggs hatch, and the larvæ undergo their complete development, within the body of the female fly. They are deposited on the ground or within birds' nests and pupate almost immediately.

The Sheep Ked (*Melophagus ovinus*).

Description and Life History.—This insect is often called the Sheep Tick, but such a name is entirely wrong. The ticks (p. 277) belong to a group of animals which are not insects, and the true ked is easily distinguished from them.

The Sheep Ked is about one-quarter of an inch in

length, brown in colour, and its very flattened body is leathery in texture. Wings are entirely absent, and the mouthparts are highly specialized for piercing skin and sucking blood. The legs are well developed and provided with claws by means of which the insect retains a firm hold in the fleece. The whole life-cycle is spent on the sheep. The adult female keds do not produce eggs, but retain and nourish the larvæ inside their bodies. As soon as they are deposited the larvæ pupate, and puparia may be found adhering to the skin and the base of the wool fibres. From these the adults emerge in about three weeks.

Nature of Damage.—Severe irritation is set up by the feeding of the keds, and serious loss of condition in the animal results.

Control Measures.—Dipping has been found to be the most effective means of treatment, and though it is more usually carried out in order to kill the Sheep Scab Mite, it incidentally destroys keds (p. 280).

CHAPTER XVII

APHANIPTERA—FLEAS

FLEAS were at one time considered to be degenerate Diptera ; they are now segregated into a distinct order, sometimes called Siphonaptera.

They are characterized by their laterally compressed bodies, their apterous condition, and by the absence of any 'neck' between head and thorax, or waist between thorax and abdomen. The head is 'stream-lined' on to the thorax: the short, stout antennæ bear three main joints, and are sunk into special grooves at the sides of the head. Eyes may or may not be present.

Mouthparts.—The mouth is ventral, and is adapted for piercing and sucking (Fig. 133). At the sides are triangular flaps of chitin, the modified maxillæ. The maxillary palps are stout and four-jointed, springing from the upper part of the maxilla. The upper lip and the mandibles take the form of long, narrow lances. The labium bears two palps, usually of five joints.

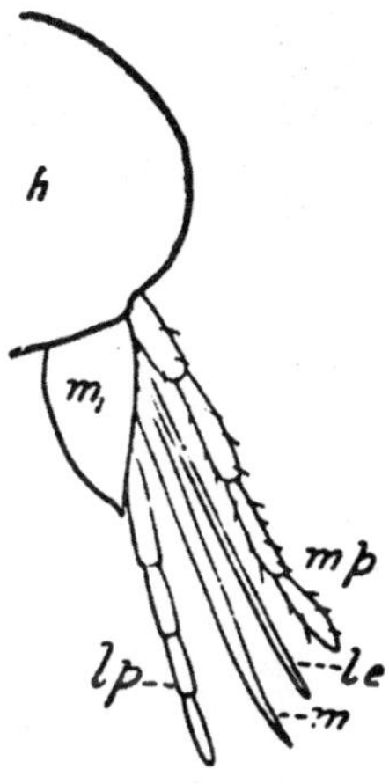

FIG. 133.—Diagram of the Mouthparts of a Flea.

h. head : *le*, labrum-epipharynx : *lp*. labial palp : *m*. mandible : *m*. maxilla : *mp*. maxillary palp.

From Imms, "Text-book of Entomology."

Wings.—Fleas have no wings, the apterous condition being secondary. The thoracic segments are all free, a rare condition amongst insects. The coxæ of the legs are well developed, and the legs increase in length as one passes from pro- to meta-thorax. The tarsus is five-jointed.

Fleas progress by jumping : some species at a single leap may cover 13 inches.

The abdomen, which in some species is capable of considerable distension, has ten segments. The whole insect is covered with spines and bristles which point backwards.

Metamorphosis.—Fleas undergo a complex metamorphosis (Fig. 134). The egg is an oval, translucent structure, visible to the naked eye. Unlike lice, fleas do not glue their eggs on to hair or feathers. The eggs hatch out in from two to ten days, giving rise to larvæ. These are whitish,

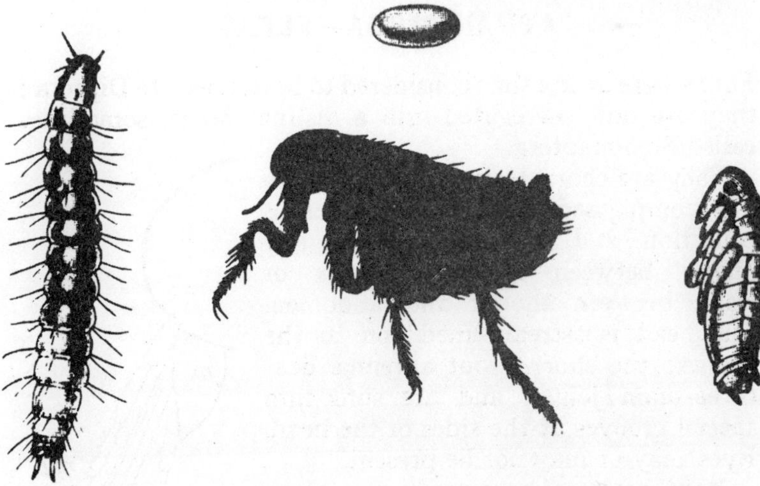

FIG. 134.—Egg, larva, pupa and imago of the common flea (*Pulex irritans*). British Museum Guide.

wormlike creatures about 4 mm. long, made up of fourteen segments including the head.

The larvæ of fleas have no eyes. Despite the absence of legs they are very active, moving by the aid of spines which cover their bodies. The larvæ themselves are non-parasitic, and cannot suck blood. They live upon organic debris, epidermal detritus, scales from feathers, etc. After three months they pupate in a cocoon.

Adult fleas are ectoparasites of mammals and birds. They suck blood after piercing the skin of their victim with their sharp modified labrum and mandibles. In piercing, the flea thrusts downwards with considerable force, elevating its abdomen and hind legs. Blood is sucked up through the grooves on the inner surfaces of the mandibles; at the same time saliva is poured out from the salivary glands into the wound. Certain yeast-like bodies extruded from these glands cause intense irritation in the wounded area.

Fleas in this country annoy the fowl, dog, cat, and rat, as well as human beings. The various rat fleas are very destructive in that they are largely responsible for the spread of Plague.

The Common Hen Flea (*Pulex gallineæ*) is the most important economic species. The dark-brown female lays her eggs or nits in dust and dirt in hen-houses. The emerging larvæ are white, with brown heads, and when mature are about one-sixth of an inch long. After feeding on refuse for two or three weeks they pupate in a brown cocoon for from ten to twenty-one days.

The adult remains in the nest or in cracks and crevices during the day and attacks the hens during the night, sucking blood and causing loss of condition due to lack of rest through irritation.

Control Measures.—Since fleas live in dirt and do not reside permanently upon the birds, obviously it is to the hen-houses and nests that most attention should be given. If the houses are thoroughly cleaned and limewashed every spring and autumn, taking particular care to force the lime into all crevices, the parasites will be destroyed. Regular attention to the cleanliness of both house and nests will prevent any accidentally introduced specimens from breeding.

Dusting birds with pyrethrum powder or sodium fluoride, as recommended for lice, will destroy any fleas present on the body. The placing of naphthalene balls in the nest does not keep fleas away, and there is evidence to show that it may have a toxic effect upon the eggs and newly hatched chicks.

CHAPTER XVIII

THE PRINCIPLES OF INSECT CONTROL

Plants and Insects in their Natural Habitats.—Before considering the relationship between cultivated crops and insects it is necessary to enquire into the state of affairs which exists under wild or natural conditions. Such wild conditions differ in many important respects from those under which crop plants are grown. If we examine the flora and insect fauna of any natural, uncultivated area of ground and see the relation which exists between plants and insects, it is at once apparent that a huge number of different species of plants flourish side by side. The smallest grasses, low-spreading types of vegetation, moderately tall herbaceous plants, bushes and trees all appear to be able to live in comfort.

The insect population is no less varied, for if we want simply to collect as many different species as possible it is to such a spot that we go. The general aspect of life in such a place changes very gradually, as the temperature, rainfall and sunshine alter with the seasons, but we may say that the flora and fauna of such an area remains practically constant from year to year. No one species or type of plant increases or diminishes in numbers, and the same species of insects are to be found just as abundantly year by year. It is clear that here we have a state of affairs which has become settled and fixed, a kind of balance, in which all the activities of nature combine to prevent any one type of plant or insect from obtaining a mastery over its fellows. The balance is the result of countless centuries of these same existing conditions, which never alter suddenly but are always undergoing a gradual and almost imperceptible change.

Some damage is always being done to plants by insects,

but the plants which we find there have survived because they can withstand a certain amount of damage and still flourish and reproduce. Those which have not these powers of resistance have died out. Similarly, the insects which damage the plants always have their numbers checked by the activities of predaceous and parasitic insects, and they are devoured too by other animals. If they tend to kill out any one type of plant they are exhausting their own food supply and suffer in consequence. The state of equilibrium thus set up is often called the Balance of Nature. Every form of life existing is to a greater or less degree dependent on every other form, and it is impossible to affect one without affecting others to some extent.

Cultivated Plants and Insects.—It is at once apparent that there is a fundamental difference between natural conditions and those under which man grows his crop plants. The mere fact of cultivation of the soil at once alters the whole state of things, for it modifies the number and kind of plants, the physical condition of the soil, its temperature, water and air content. It is in fact a sudden upset of the natural balance, and has a most profound effect. Furthermore, man supplies food to the plant and so encourages its growth, but perhaps the greatest alteration is in the actual types of plants. We find nothing approaching our mangolds under wild conditions, but the little plant *Beta maritima*, which grows on the seashore, is said to be the original parent of them. Our cereals are derived from wild grasses and our apples from wild apples, but the forms which we cultivate are vastly different from those wild ancestors. The difference is not only in appearance but in the inherent properties of the plants. They are types which flourish under conditions of cultivation and only under those conditions, for their natural powers of resistance and recuperation are often very limited. In Nature also we do not find a great aggregation of the same species of plants, such as man establishes, in the form of fields of cereals or roots, and orchards containing only apples or plums. Man can only exist because he has power to alter natural conditions in a very great degree: he has established what is, in effect, an artificial balance, which he is always striving to maintain, and which is always tending to revert back to the natural one. He can control a great many factors affecting his crop plants, but he has also to take account of the fact that animals, and insects in particular, are also dependent on plants. Where a great number of plants of the same sort are congregated, the insects which feed on them must congregate also, and the business of overcoming the damage they do is one of man's activities in maintaining the balance, which he endeavours to weight in his favour.

The Knowledge necessary to devise Control Measures.—To most people insects are more or less of a mystery. They suddenly appear and as suddenly vanish into the unknown again, though it is quite obvious if a little thought be given to the matter that some form or other of the creature must always be living somewhere. Yet few associate caterpillars with butterflies or moths, and grubs in the soil with beetles. Where certain caterpillars are doing severe damage to trees in the spring, it is not always a simple matter to convince people that those caterpillars arise from eggs laid by moths several months before. If one were to claim that he could prevent caterpillars damaging trees in May by catching moths in the previous December, his assertion might be regarded with some doubt. The majority of people would prefer to wait until the caterpillars appear before doing anything to stop the damage, and this is largely the case because they do not associate one stage of the insect with another. Their control measures take one form only, namely, the actual destruction of the harmful stage, usually a difficult, expensive, and often quite impracticable task. A complete knowledge of the insect's life history and the habits of its various stages opens up many more possibilities. We can often destroy an insect much more conveniently in some of its harmless stages than in the harmful ones, and it does not matter at what point we break the life-cycle, we shall achieve the desired object in eliminating the destructive stage. As an example of this, the use of egg-killing winter washes on fruit trees may be given. By the use of such washes it is possible to control such insects as aphides and apple-sucker much more effectively in the winter than by a summer spraying after the eggs have hatched. Again, if we anticipate an attack and take steps beforehand which make conditions adverse to the insect when it appears we shall also prevent damage from being done. Measures of this sort are chiefly cultural in nature and consist in doing the right operation at the right time, often long before an attack commences. For example, frit fly attacks spring-sown oats about the middle of May, and once an attack is in progress there is no remedy for it. But the seed-bed is prepared and those oats are sown in late February or early March, and the amount of growth they have made by the middle of May largely determines whether or not they will be attacked. The nature of the seed-bed and the time of sowing have a most marked effect on this, but the necessary knowledge must be at the command of the farmer, and he must exercise foresight and anticipate the attack if he is to avoid it.

All methods of control can be classified under three headings, but the actual nature of the measures in each type will necessarily differ in detail owing to the fact that they have to be adapted to little differences in the habits of each insect.

1. **Direct control measures,** in which the aim is actually to kill the insect by the application of some toxic substance, or by mechanical means.
2. **Indirect control measures,** which consist chiefly of cultural operations. It is found that certain kinds of cultivations tend to favour the plant's growth and at the same time to make conditions unfavourable to the insect concerned.
3. **Biological control measures,** which utilize enemies and diseases of the pests in keeping down their numbers.

1. Direct Control Measures.

These are chiefly applicable where crops have a high value, such as fruit or market garden crops, because they are naturally expensive both from the point of view of materials and labour. In a few instances, however, they are applied on an agricultural scale. The main kinds of direct measures are as follows :—

Sprays and washes.	Fumigants.
Poison baits.	Dips and Dressings.
Traps.	Deterrents.

Sprays and Washes.—These may be either in the form of powder (dry sprays) or liquids, and are applied by means of a special machine. They fall naturally into three classes:

Stomach poisons.
Contact poisons.
Egg-killing washes.

Stomach poisons are usually arsenical compounds. They can only be used against insects and larvæ which obtain their food by chewing pieces out of the plant. The insect's food is covered with a very fine layer of the poison, which is only effective when swallowed. The chief use of these poisons is on fruit trees and bushes, against leaf-eating larvæ and adult beetles.

Contact poisons differ essentially from stomach poisons in that they are only effective when they actually touch the insect or larva. Nicotine or nicotine sulphate are the two chief contact poisons, but derris extract has well-marked properties of the same nature. Pyrethrum and hellebore powders are used in some circumstances. Soft soap solution, especially when mixed with quassia extract, acts as a contact poison, but the effect is probably due in part to the soap film blocking the breathing system of the insect. It is important to realize in using these substances that they act only when they actually come into contact with the insect or larva. Sucking insects, chiefly aphides and capsid bugs, are controlled by contact poisons. Stomach poisons are useless against them, because these insects obtain their food from the sub-epidermal cells of the plant and do not actually devour leaves, etc. Certain sulphur compounds appear to act as contact poisons, and are particularly employed against various species of mites. Dry applications of lime, soot, sodium fluoride, and especially copper sulphate, are used against slugs.

Egg-killing washes.—The old practice of limewashing fruit trees in winter or early spring was effective to some extent, because insect eggs on the trees were sealed up and prevented from hatching. More recently new egg-killing washes have been introduced, which are obtained from coal-tar distillation. Applied during the winter, when the trees are in a dormant condition, they are capable of killing insect eggs exposed on the trees. The eggs of aphides and apple-sucker are effectively destroyed, and some 'kill' is obtained on the eggs of the winter moth and apple capsid bug. On no account may these washes be employed after the buds are showing signs of swelling in early spring.

Poison Baits.—These consist essentially of artificial attractive baits, mixed with some form of stomach poison. Poisoned syrup has been used against certain flies, using sodium arsenate as the toxic substance. Bran, mixed with a small quantity of Paris Green, is employed as a poison bait for leather-jackets, cutworms and slugs. Slices of potatoes, beetroot and carrots, soaked in a mixture of Paris Green and water, and buried just beneath the soil, act as poisoned baits for many soil-living larvæ, and are used on a small scale.

Traps.—A great variety of devices may be regarded as traps. Elaborate machines have been made for catching such insects as flea beetles in the field, and there are many other forms of traps for various insects, down to the tile or board laid on the surface of the soil to provide an artificial shelter for slugs. Sticky fly-bands are in the nature of traps, as are grease-bands on fruit trees. Some insects, particularly moths, are attracted to lights at night, and light traps have been made which are based on this attraction. In a few cases substances which actually attract insects are utilized as a kind of bait. The sense of smell is highly developed in many insects, and they are undoubtedly able to detect small quantities of an attractive substance from great distances and go towards the source of it just as others go towards light.

Fumigants.—Gases evolved either by direct chemical action or by the evaporation of a volatile liquid are spoken of as fumigants. They can only be effective in glass-houses, warehouses and buildings in general, or in closed spaces such as canvas tents, etc. In any place where a sufficient concentration of gas can be obtained this method is very effective in killing all active stages of insects, but eggs are to a large extent unaffected. Soil sterilization with steam or with soil insecticides, such as carbon disulphide, may be looked upon as fumigation, since the small air spaces in the soil do not permit of rapid diffusion of the gas.

Dips and Dressings.—It is common to apply these terms to the treatment of animals for various external parasites. The substances used are almost entirely contact insecticides, and their action is the same as that of contact washes used on fruit trees.

Deterrents.—Just as insects are attracted to various substances by their sense of smell, so they may be repelled by other substances, the smell of which they dislike. Such substances are known as deterrents or repellents, but our knowledge of the sense reaction of insects is very limited, and little use is made of this principle in practice for preventing insect attacks.

2. **Indirect Control Measures.**

Measures of this nature are difficult to classify, since they act in a very complex fashion. They may be defined as operations affecting the conditions under which crops are grown, producing a favourable result on the plants and discouraging the insects. No insecticidal substance is applied; no insects are directly killed, though they may be either driven away or find conditions so adverse that their numbers become negligible.

We have already seen that the conditions under which crop plants are grown are very different from those which exist in nature. Cultivation has made possible the growing of less hardy but more productive types of food plants, and enables rapid changes in soil conditions to be carried out. Changes of this nature profoundly affect insect life. A field which carries, say, a cruciferous crop during one year and in the succeeding year grows barley, cannot maintain a settled insect population. We see at once that a rotation of crops is in itself the foundation of the indirect control of insects. Without a doubt rotations became established partly because they had this effect and resulted in better and more certain crops. Only certain insects will survive these conditions, and they will be either of the omnivorous type, *e.g.*, wireworms, or insects which are able to exist on weeds of the same botanical order as the cultivated plants. Two botanical orders are perhaps more important than others in this respect, the Gramineæ and the Cruciferæ. Grasses, of course, are indispensable on the farm, but they support many flies which do damage to cereals, while cruciferous weeds are so abundant that such insects as flea beetles find ample breeding and feeding grounds by the roadside. Rotation has played its part in the suppression of insects and reduced the numbers of serious pests to a very few species. These species, if they are to be controlled, must be dealt with by more specific methods, which a study of their life history and habits indicates. How may conditions be still further modified so that in one and the same operation the insects are discouraged and the plants helped on ?

An examination of damage to crops by insects brings out one prominent feature, namely, that there are susceptible stages in a plant's growth. If it gets through these critical periods quickly and in good time, the damage is usually small: if the growth is delayed, damage results. The most critical time is that in which the plant is becoming established just after germination, a period which is conveniently called the 'seedling stage.' With few exceptions, plants once firmly rooted and in full growth suffer little, for they usually outgrow an attack. The exceptions are found in those plants which are damaged between the flowering stage and harvest, but such cases are relatively few. The seedling stage is all-important, and the plant must be given every inducement to grow through it as quickly as possible. Two factors, apart from weather, have a profound effect, namely, the supply of plant food and the texture of the seed-bed. Manuring is perhaps the more easily controlled factor of the two. A shortage of any constituent can be ascertained beforehand and rectified fairly quickly. Adequate supplies of humus in the soil affect the availability of the plant food to a large extent, and the use of good farmyard manure is vitally important for this as for many other reasons. Readily available supplies of phosphates help the plant through its 'seedling' condition, and light top-dressings of nitrogenous fertilizers may be very beneficial if used with discretion. They make for a more vigorous growth, however, rather than for advancing the actual stage of growth, and are more effective in inducing the plant to make good any loss of leaf area than in any other way.

Given the requisite supply of plant food, then, the physical texture of the soil is pre-eminently important. That texture controls to a large extent the amount of water which the seedling plant gets, and through it the supply of food. It may or may not be conducive to rapid and proper rooting of the plants, in fact it controls the start which the crop gets, and a good start is proverbially more than half the battle. Small seeds, with little reserve food supply, need the very best conditions if they are to make a favourable growth in the seedling stage, and the absolute necessity for obtaining the very best seed-bed cannot be too strongly emphasized. The texture of the seed-bed, and to some extent the manuring, must be controlled with due regard to the type of soil with which one is dealing, but 'roughness' in the seed-bed is a disadvantage. Autumn-sown cereals, especially on heavy land, do not need so good a seed-bed as spring-sown crops, since they become established at a time when they are not liable to attacks. They are not subject to drought, and heavy winter rains would have an adverse effect if the soil were in a finely divided state. Spring sowing, however, is done at a most critical time, when skill in cultivation counts. A large part of that skill comes from foresight, for if, when cultivating, the farmer anticipates a possible attack, he will have more inducement to take the extra trouble necessary. The preparation of a seed-bed must be regarded as an insurance, and the assumption that there will be no insect attack on the crop means that the farmer is taking an unjustifiable risk. A rather more careful handling of the soil, an extra cultivation or two, though costing a little more, constitute a good investment, for a partial crop means almost as much subsequent cultivation and labour in harvest and threshing. It leaves a dirty field and no profit.

The time of sowing has an influence also on the attacks by insects, because it is a factor which assists the plant to become established in good time. Especially in the case of spring-sown cereals, early sowing has a very beneficial effect. It must be pointed out, however, that very early sowing on a bad seed-bed may be no more advantageous than rather later sowing on a better tilth. The important combination is to secure a good seed-bed and to make every effort to secure it as early as possible. Firmness of the soil is another quality closely connected with the rapid establishment of the plant and its unhindered growth. It can only be obtained when the tilth is evenly worked and fine in texture throughout its entire depth. Large clods beneath the surface retain large air spaces between them, and seedlings are unable to root properly under these conditions, however fine the top layer may be, because the soil cannot be made solid below. Firmness not only makes for a continual supply of soil water and assists good rooting, but it also restricts the movement of soil larvæ. Once the seedling has become well established, subsequent cultivations carried out for cleaning the land all benefit the plant, but, apart from exposing some insects to the attacks of birds, they cannot be looked upon as an actual controlling influence. In certain cases, however, for example in connection with Turnip Flea Beetle attacks, a considerable amount of good may be done by maintaining a fine dusty surface tilth.

Although hardly to be considered as an act of cultivation, the process of general cleaning up on the farm and in the orchard or garden is extremely beneficial. Many insects pass over the winter in the adult stage, hibernating in any suitable quarters. Stack bottoms, heaps of hedge brushings, dry grass refuse and cleanings from ditches, etc., harbour innumerable insects. If properly handled they act as traps, and if burned before spring comes they may be a means of destroying a great many pests. On the other hand, as soon as the earliest warm days come, insects leave these quarters and search for their food plants. It is difficult to estimate how much good is actually done by the destruction of such rubbish, but it is certainly very considerable. Closely connected with this is the suppression of weeds, and the elimination of untended weedy corners and waste places from which insects migrate to crop plants.

3. **Biological Control.**

Although the majority of insects are plant-feeders, there is a considerable number which are carnivorous. The Carabidæ and Coccinellidæ among the beetles are mainly carnivorous in both adult and larval stages. The larvæ of certain flies, especially in the families Syrphidæ and Tachinidæ, prey upon insects. Some of the Rhynchota (Heteroptera) and various mites (Acari) are predaceous, while the larva of the Lace Wing Fly (*Chrysopa*) is a well-known aphis feeder. In the order Hymenoptera there is one whole group known as the Parasitica, insects which oviposit in or upon other larvæ and eggs. The larvæ of these parasitic insects feed upon and eventually destroy the others and so serve as a natural check upon their increase. This true parasitism is confined to the Hymenoptera and certain families in the Diptera. The yellow cocoons spun by the parasitic larvæ are frequently found surrounding the dead bodies of lepidopterous caterpillars. The larvæ of the parasites may themselves be parasitised also, a phenomenon known as hyper-parasitism, which is by no means uncommon.

Attempts have been made to utilize these parasites for the suppression of some insect pests. It frequently happens that an insect which is not particularly numerous in its native country, suddenly reproduces enormously if it gains admission to another country. Thus the Gipsy Moth, which is not regarded as a great pest in Europe, has become almost a plague in the United States of America since it was accidentally introduced into that country. This sudden increase in numbers is often largely due to the fact that the insect finds itself in an area where it has no natural parasites to check its rate of reproduction. If the parasites are introduced the balance may be established again, though since many other factors are at work it is often found that this method does not give the results hoped for. In England we find that most aphides are heavily parasitised as well as preyed upon by ladybirds, etc., and were it not for these natural checks it is doubtful if man would be able to suppress the damage which aphides cause to crops. Little has been done, however, to exploit parasites in this country, except in the case of Woolly Aphis. This insect is remarkably free from parasitism, but one small Chaleid, *Aphelinus mali*, does destroy it in America. It has been introduced into England, and it remains to be seen if it will live and be effective under these climatic conditions.

Not only do some insects exercise a control upon others, but some of the mites already mentioned, and spiders are also carnivorous. Many mammals, such as the mole and bat, are insectivorous, while numerous birds do much good by searching for and feeding upon insects in all their stages. Apart from the protection of useful birds, however, little has been done to exploit these natural methods of control.

CHAPTER XIX

INSECTICIDES: THEIR COMPOSITION AND APPLICATION

It has already been shown that insects can be controlled by a variety of methods which have been classified as *Direct, Indirect,* and *Biological.* The Direct methods aim at the actual destruction of the insects concerned and may be either of a mechanical nature or involve the use of poisons. The problem of destroying insects by means of these poisonous insecticides is by no means so simple as it might appear. Many considerations arise, and if these are not taken into account the work may be not only futile but definitely harmful. It is therefore necessary to enquire into the conditions which make for the successful use of insecticides, and also to understand the chemical and physical composition of the substances employed.

Insects, as we know, all undergo metamorphosis to a greater or less degree, and we have seen that only two stages are destructive because there are only two feeding and active stages, the adult and the larva (or nymph). Since the larva (or nymph) is the growing stage it naturally feeds more than the adult and is almost invariably the destructive stage. It does not matter, however, whether we destroy this stage or not, for if we can break the life-cycle at any point, the larva will be prevented from appearing, and it is often more convenient to concentrate on one of the dormant stages rather than the larva itself. Adult insects, as a rule, do not feed to any very great extent, but constitute primarily the reproducing and dispersing stage. The first essential, therefore, is to discover at which stage the insect is most vulnerable and can be most readily destroyed, a fact which can only be established by consideration of the complete life history and habits.

Again, poisons which are effective against some insects are useless against others, and even when the right substance has been found the most effective method and time for applying it have also to be discovered.

The effect of the application of poisons on the plants themselves is another vital consideration. If the substances are not correctly prepared or in the right form, great damage may result. The physical state of a liquid is perhaps even more important in this respect than its chemical composition. When a soluble salt, such as copper sulphate, is dissolved in water and the solution sprayed on to green leaves, as in charlock-spraying, severe scorching results and the plant may be killed out. Such scorching may be due either to plasmolysis or to the caustic and poisonous properties of the soluble salt, and it also occurs if the finely divided salt is applied in solid form. It is therefore clear that, however effective soluble salts might be as insect poisons, their use is at once ruled out on this account. The alternatives are either to use an insoluble substance on green leaves or to employ substances which do not cause scorching. Now many salts, insoluble in water, act as poisons when eaten, and if these can be applied satisfactorily they may reasonably be expected to be effective. Apart from using them in the form of powders or dusts, they can be applied in liquid form as *suspensions*. A suspension consists essentially of very fine solid particles dispersed throughout the volume of water, but no actual solution takes place. The more finely divided the powder, the better will be the suspension and the less tendency there will be for it to settle. Such a suspension, sprayed on foliage, leaves a very fine deposit of the solid adhering to it when the water evaporates, and no scorching results. Certain very poisonous substances, notably lead arsenate, are commonly used in this way. Similarly, certain oils, such as common paraffin or kerosene, cause very severe scorching of foliage in their raw state. They can, however, be made up in the form of emulsions, in which the oil is broken up in very tiny droplets and dispersed throughout the water, much in the same way as a suspension of solid particles. In such a state many oils can be safely applied without scorch resulting, but precautions have to be taken to prevent the emulsion from breaking down.

On the other hand, there are poisons like nicotine which mix readily with water in almost any proportions and cause no scorching at all.

Insecticides are employed in three forms—Powders, Liquids, and Gases. When powders are used the operation is spoken of as dry spraying or dusting; the application of liquids as spraying or washing, and the use of gases is known as fumigation. Poisons act in one of two ways. Certain compounds are effective only when eaten and when they gain access to the digestive fluids. These are known as *Stomach Poisons*, and the toxic principle is usually arsenic. On the other hand, substances like nicotine rapidly act upon the nervous system when insects are wetted by them, and they bring about paralysis and death. Such poisons are known as *Contact Poisons*, and are chiefly plant extracts. Gases, acting through the breathing system, seem to act in a similar manner to contact poisons.

FORMS OF INSECTICIDES

Powders.—Both stomach and contact poisons are prepared in powder form, and the essential of all such powders is that they shall be in a very finely divided state. They are applied either by hand or spread by means of a wind-blast from bellows, which are worked by hand or incorporated in more or less elaborate dusting machines. On a large scale, in England, dusting is but little practised, since climatic conditions are generally adverse. In America the use of powder insecticides is common, and over large areas the powders have been applied from aeroplanes. If powders are to be used successfully in the field, a still atmosphere is almost essential, for if there is much wind the plants cannot be adequately covered and the operator may be seriously inconvenienced. Heavy rainfall soon after application tends to wash off the powder to a large extent and render it useless. Insecticidal powders are therefore only used to any extent in this country inside buildings or as dressings against animal parasites, but there are a few cases in which they find application on a field scale.

Liquids.—The application of poisons in liquid form is by far the most usual method, and such liquids are commonly applied by means of spraying machinery. A pump, operated either by hand or by means of a small petrol engine, is used to obtain the necessary air pressure, which drives the liquid through a nozzle, splitting it into fine drops and spreading it in a broad cone of spray.

Insecticides in liquid form fall into three types:

1. Suspensions of insoluble salts in water—chiefly stomach poisons.
2. Mixtures of substances, such as nicotine, with water—chiefly contact poisons.
3. Emulsions of oils—contact poisons or egg-killing washes.

Not only are spray fluids used on foliage, but certain emulsions, known as Tar Oil or Tar Distillate washes, are now used as winter egg-killing washes. They scorch badly if used on foliage, but have the power of killing many insect eggs and so are used in winter on fruit trees when they are in a dormant state.

Insecticidal dips for animals (*e.g.*, sheep dips) are usually in the nature of Contact Insecticides.

Gases.—The dangers attending the use of poison gases in the open air makes the use of insecticides in this form almost impossible in the field. Consequently fumigation is only employed in buildings such as glass-houses, mills, stores, etc., but in America, with some valuable crops, canvas tents are erected which enclose individual trees, and these are then subjected to fumigation. The gases used are chiefly hydrocyanic acid gas, tetrachlorethane, nicotine vapour, and carbon disulphide. These are produced in the building to be fumigated, the place having previously been rendered as air-tight as possible. Under good conditions fumigation is fatal at all active stages of insect life, but eggs are unaffected, and the operation has often to be repeated two or three times at intervals in order to destroy the young stages as they hatch.

Stomach Poisons.

Lead Arsenate.—This substance is a white powder, almost

insoluble in water. It is on the market in two forms, one as a dry powder and the other as a paste with water. It is a very powerful insecticide, since both lead and arsenic are poisons. As a standard insecticide against most leaf-eating larvæ and some adult beetles, it is used by fruit growers in the following proportions:

Lead Arsenate, (powder) 2½ lbs. or (paste) 4 lbs.
Water, 100 gallons.

The lead arsenate is first made into a thin cream with a few pints of water, very thoroughly mixed to break up any lumps, and then added to the bulk of the water, with continual stirring. Good brands of lead arsenate remain in suspension without constant stirring, but if there is any tendency for the solid particles to settle out the liquid must be stirred frequently.

Sodium Arsenate.—Unlike lead arsenate this sodium compound is very soluble, and is the substance commonly used to poison syrup which is exposed for flies to feed upon. The following formula has been used against the Onion Fly (*Hylemyia antiqua*):

Sodium Arsenate . .	¼ oz.
Treacle	¼ pint.
Water	1 gallon.

Paris Green.—This compound was formerly used as a paint colour, but is now often employed as an insecticide. It is a brilliant green, very fine powder, and is a copper aceto-arsenite. To a considerable extent it is soluble in water, but has been used in the past as a poison for spraying on foliage. Its tendency to scorch, however, led to its replacement by lead arsenate, and it now finds a use in the making of poison baits. Such baits have been used with considerable success on a field scale and in market gardens against leather-jackets, cutworms and slugs. The formula for making a suitable poison bait is as follows:

Paris Green . .	1 lb.
Bran . . .	30-40 lbs.

The bran is first slightly damped with about a gallon of water sweetened with sugar or treacle. Only just sufficient water should be used to make the bran damp, and the flakes should not adhere together and become lumpy. The Paris Green is mixed, in small quantities at a time, with the bran, which gradually acquires a faint green tinge. Too much Paris Green should not be used as this tends to make the bait unpalatable. Such a bait, sown broadcast on infested land, gives an excellent control of leather-jackets, but for the maximum success two or three warm rainless days and nights are necessary.

Derris extracts (see Contact Poisons).

Contact Poisons.

Nicotine.—Certain plants, especially some belonging to the Natural Order Solanaceæ, contain substances known as alkaloids. These are highly poisonous, and the best known one is that obtained from the tobacco plant. Commercial nicotine is a dark brown, very unpleasant-smelling liquid which mixes readily with water and unites with acids, forming salts such as nicotine sulphate, which are also poisonous. Although, like most poisons of this group, the effect is most marked when the substance is swallowed, yet in the open this characteristic is soon lost, and leaves sprayed with nicotine are poisoned only for a very short time. This alkaloid, however, possesses the very useful property of being poisonous when it comes into contact with soft-bodied insects and larvæ, a characteristic not exhibited by typical stomach poisons. It is this property which makes nicotine such a valuable poison for use against insects with piercing and sucking mouthparts, for many of these cannot otherwise be readily controlled. It should be realized, however, that in order to kill the insects, they must be thoroughly wetted by the fluid.

A common formula for making a spray fluid of this nature is:

Nicotine, 6-12 fluid ozs., or,
Nicotine sulphate, 15-30 fluid ozs.
Water, 100 gallons.
Soft soap, 2-3 lbs.

The soft soap is first dissolved in a few gallons of hot water and then added to the bulk of water. Finally, the nicotine is added immediately before application, and the whole well stirred.

Soft Soap.—This substance by itself is often used on delicate plants against such insects as aphides. It cannot be regarded as a poison, but appears to be effective because it forms a film over the spiracles of the insects, and brings about their death by asphyxiation. It is included in many other contact sprays owing to its action as a 'spreader.' Where possible, soft water is always used in making up spray fluids, but pure water, sprayed on to foliage, always tends to remain in the form of drops. If soft-soap solution be added to the water it will be noticed that the drops tend to cover a much larger area and become flatter, due to a lowering of the surface tension of the water. This increases the spreading or covering power of a spray, making it more economical in use, and enabling the liquid to penetrate into small crevices which it would not otherwise enter. As a rule sufficient soft soap should first be used to destroy the 'hardness' of the water, and then some extra added in order to improve the spreading power.

Quassia.—This is an extract from the wood of certain tropical trees. It is purchased as quassia chips, which are boiled in water, and the extract has an exceedingly bitter taste. Its insecticidal value is probably small, but it was formerly mixed with soft-soap solution and used against aphides and such insects.

Derris Extracts.—Extracts from the roots of the tropical plant *Derris elliptica*, or Tuba-root, have long been known as fish poisons. They are found to affect cold-blooded animals, including insects, but are not poisonous to warm-blooded animals. The high price of nicotine has led to the exploitation of these extracts as insecticides, and at present most of the non-poisonous insecticides on the market contain this substance. The finely powdered parts of the plant are stirred up with water and the mixture applied at once. The addition of soft soap is often recommended, but in some preparations is incorporated in the powder. Not only are derris extracts contact poisons, but, for a time at any rate, they act as stomach poisons when leaves which have been sprayed with them are eaten.

Since derris-containing insecticides are at present all proprietary articles, they should be used at strengths recommended by the makers.

Non-poisonous substances of this nature possess the great advantage of being suitable for use on vegetables and bush fruit where spraying is often needed just before the produce is fit for use.

Lime Sulphur.—The chemical interaction of lime and sulphur produces a complex substance, chiefly sulphides and poly-sulphides of calcium, which is known commercially as lime sulphur. This is a reddish liquid with a specific gravity of about 1.3. Diluted with water it is often used as a fruit-tree spray, particularly as a winter wash, though its insecticidal action is less pronounced than its effect on fungi. In winter it is used at a strength of 1 part of lime sulphur (sp. gr. 1.3) in 20 parts of water. If used on foliage there is considerable risk of scorching, but strengths of 1 in 60, and, on susceptible plants, 1 in 100 have been used. Lime sulphur is most effective against mites (Acari) and is now recommended for use against the 'big bud' mite of black currants. Certain varieties are liable to slight scorching, but this danger is greatly lessened if spraying is done in mild and not frosty weather. It is also commonly used as a dip against sheep-scab mite.

Oil Emulsions.—Paraffin oil in the form of an emulsion is sometimes used as a winter-spray fluid. It is unsafe to use it on foliage, since there is considerable danger from scorching, but it is sometimes employed as a late summer spray after fruit has been gathered. At this period many fruit tree insects, such as aphides, are about to lay their eggs and are destroyed before doing so. The tree may be defoliated prematurely, but this has little ill effect.

The emulsion should be carefully made, and the following formula is suitable:

Paraffin (kerosene)	1 pint.
Soft soap	1 lb.
Water	10 gallons.

Miscible Oils.—These are in the nature of light-grade lubricating oils prepared in an emulsified form. They have not been used widely in England at present, but appear to be efficient in action and practically harmless to foliage, although some cases of damage have been recorded.

Tar Oils.—A recent addition to the list of insecticides, these oil emulsions are of outstanding value, since they are capable of destroying the eggs of many insects. They are very complex substances from a chemical standpoint, and are fatal to foliage. Their use is confined to winter washing of trees in a dormant condition, at which period they destroy the eggs of aphides and apple-sucker, and to a large extent those of the winter moths. It is regarded as unsafe to spray plums with them later than the end of January, and apples later than the end of February. The strengths recommended by the makers should be employed. There are many different brands of these washes now on the market, and some appear to give definitely better results than others.

Pyrethrum Powder.—This powder is made from the flower heads of a Japanese pyrethrum, and is effective against lice, cockroaches, etc. Its use is confined to animals and buildings, and it is not largely employed. A similar substance is Hellebore powder, also a plant product.

Copper Sulphate.—Although it cannot be looked upon strictly as an insecticide, copper sulphate has a marked poisonous action upon slugs, which are pests of great importance. Many substances, such as sodium fluoride and powdered alum, have been used for this purpose, but they are not so toxic. A very tiny crystal of copper sulphate touching the back of a slug is sufficient to destroy it, and the following mixture has been found efficacious in dealing with slugs, particularly in cereal crops :

Copper sulphate (finely ground) . . .	6 lbs.	3 cwt. of the mixture per acre.
Commercial kainit .	120 ,,	

The kainit, or other suitable substance, acts simply as a carrier for distributing the copper sulphate, which by itself could not possibly be applied at the rate of about 18 lbs. per acre. This mixture is broadcast on a fine mild evening when the slugs are on the surface of the ground. It is important that application should be made at the right time, since only those slugs actually touched are killed. Evenness of distribution is an important factor in obtaining the maximum success.

Fumigants.

Hydrocyanic Acid Gas (HCN).—This is a colourless gas with a characteristic smell, and it is extremely poisonous, as are the cyanides used in its preparation. It is usually evolved either by action of sulphuric acid on sodium cyanide, in suitable vessels, or by the action of atmospheric carbon dioxide and water on calcium cyanide. The reactions take place as shown:

$$2NaCN + H_2SO_4 = Na_2SO_4 + 2HCN$$
$$Ca(CN)_2 + H_2O + CO_2 = CaCO_3 + 2HCN.$$

Hydrocyanic acid gas is soluble in water, and therefore there is some danger of scorching if much water is present and if the operation is carried out in strong sunlight. In order to avoid these dangers it is usual to carry out the fumigation of glasshouses at night, and to have the plants rather dry at the roots. A temperature between 40° F. and 60° F. is best, since damage may occur outside these limits. The house must be thoroughly ventilated before men are allowed inside again. White Fly (*Aleyrodes*), an insect very difficult to deal with by other means, is controlled by fumigation, and the following quantities are commonly used per 1000 cubic feet of house:

Sodium cyanide	$\frac{1}{4}$ oz.
Sulphuric acid	$\frac{1}{2}$ oz.
Water	2 oz.

Tetrachlorethane ($C_2H_2Cl_4$).—The use of very powerful poisons such as cyanides is attended by some considerable risk, and various less dangerous substitutes have been tried as fumigants. Of these tetrachlorethane has been almost as successful as cyanide. It is a volatile liquid which is poured down the centre of the house at the rate of about half a pint per 1000 cubic feet of space, and the house is then closed.

Carbon Disulphide (CS_2).—This is a volatile liquid, giving off a very heavy vapour at ordinary temperatures. It is extremely poisonous, evil-smelling and inflammable. The danger of employing it in buildings has always been against its widespread use, and it finds its chief commercial application now in soil fumigation, chiefly against root colonies of woolly aphis on apple-trees. The tiny spaces between soil particles enable a high gas concentration to be maintained, and all active forms of soil life are destroyed.

Sulphur Dioxide (SO_2).—Owing to its extreme solubility and poor penetrating power this gas is little used. It can only be employed as a kind of cleaning-up treatment in a glasshouse when the crop is gathered, and the house about to be cleared. The usual method of preparation is to burn lumps of sulphur on the floor of the house.

Nicotine.—Liquid nicotine can be readily vaporized by heating it in a suitable vessel over a spirit-lamp. The vapour is extremely poisonous, and every precaution must be taken by the person using this treatment. It is little used now on a commercial scale.

Steam.—The super-heating of buildings by steam has been found effective in destroying insects, particularly in corn stores, warehouses, etc.

THE USE OF INSECTICIDES

By far the greatest quantity of insecticidal material is used by fruit-growers. Fruit-tree spraying is as necessary a part of the routine of the modern commercial plantation as pruning and such-like operations. The machinery, whether operated by hand or by engine-power, should be adequate to cover the work rapidly, since it is often necessary to complete a spraying with, say, nicotine in a few days, if it is to be successful. It is only at certain periods that many insects are vulnerable, and to apply an insecticide at any other time is largely waste of time and material. Thus, taking the leaf-curling plum aphis as an example, it is known that very shortly after the female aphides hatch in spring their feeding causes a curling of the leaves. Under the very best conditions it is practically impossible to force a fluid inside leaves thus curled, so that the operation must be done before any curling takes place.

There exists in the life-history of most insects some vulnerable stage or phase when they can be dealt with successfully; but since each insect needs almost a specialized treatment it is difficult to lay down any general principles. Certain principles do hold, however, governing the type of insecticide used, and each type needs to be applied in a different fashion.

Stomach Poisons.—The object of using poisons of this type is to put a very fine coating of the substance over the insect's food. In order to accomplish this most economically it is clear that the whole surface must not be thoroughly wetted, since if it is, the tiny solid particles in suspension will tend to accumulate at the lowest points and run off. If just sufficient liquid could be applied to give a continuous film, without any surplus, this would be the ideal, but in practice it is found best just to cover the surface with a large number of tiny droplets, which on evaporation leave an almost continuous coating of poison. In order that this may be best accomplished, it is necessary for the spray fluid to be very finely divided indeed, so that it is applied in the form of a mist or cloud. The surface to be sprayed is not wetted sufficiently to make it drip.

All insects and larvæ with mandibulate mouthparts can, as a general rule, be dealt with by means of a stomach poison. Leaf-eating weevils, caterpillars of the winter moth, and many others, are typical of the kind of insects against which such poisons are effective.

Contact Poisons.—The effective use of a contact poison requires certain conditions exactly opposite to those which are most advantageous in using stomach poisons. Contact poisons are effective only for a short time after application, and to kill, they must hit the insect. It is indeed the insects themselves which are sprayed, and not the insects' food, as is the case with a stomach poison. Thus a high pressure is necessary for a contact spray in order to force the fluid into curled leaves, flower trusses and crevices of bark, etc., and the actual spray, instead of being in the form of a fine mist, consists of relatively large drops. Moreover, it is desirable to cover as much surface as possible by the liquid, so that a spreading agent like soft soap is commonly employed.

All those insects with piercing and sucking mouthparts, particularly members of the order Rhynchota, have to be controlled largely by contact poisons. Compounds such as lime sulphur appear to act in a similar manner on the mites (Acari).

Egg-killing washes must be applied in a similar manner to contact washes, since a very thorough covering and good power of penetrating into crevices is necessary. Such washes may be used on all fruit trees and bushes during the dormant season, but on no account should trees be sprayed after the buds show any marked development, or they may be badly scorched, and in the case of plums a large proportion of the blossom buds are killed outright and drop off. Certain varieties are always more forward in growth than others, and there is also a seasonal variation, but, in general, plums may safely be sprayed up to the end of January, and apples up to the end of February.

During the winter months many insects are present on fruit trees in the egg stage, particularly aphides, apple sucker, winter moths and apple capsid bug. An infestation by these insects in the spring necessitates a costly and difficult series of spraying operations, and a more effective control can be secured if tar distillate washes are used and the eggs destroyed.

In the case of aphides and apple sucker it is possible to reduce the hatch of eggs almost to *nil*, but only a partial control of winter moth and capsid bug is obtained as a rule.

Since the objective of a winter spraying is to hit the eggs of the insects, the liquid must be applied with as much force as possible, and the whole of the tree thoroughly wetted, especially the small wood- and fruit-spurs. Green crops—*e.g.*, brussels sprouts—beneath fruit trees will be badly scorched by these washes.

Fumigants.—The whole success of fumigation depends upon maintaining a sufficiently high gas concentration throughout the building. This cannot be achieved if there are any appreciable leaks due to badly fitting woodwork, broken glass, etc., and the success of the operation will largely depend on eliminating all such ways of escape for the gas. A temperature of between 40° F. and 60° F. should be maintained, and the operation is best done at night, since there is then less risk of the plants being scorched. Certain plants, however, are much more susceptible to scorch than others, and in a house containing many different kinds of plants it is difficult to adjust the dosage so as to be safe for them all.

It is important that the operator shall leave the house as soon as the gas is evolved. In the case of hydrocyanic acid gas the operation is often done from outside by arranging some mechanism by which the sulphuric acid can be poured over the cyanide, but if carried out rapidly in the ordinary way there is little danger. The period of fumigation usually lasts about twelve hours, and before any persons are allowed to remain in the house it should be thrown open and very thoroughly ventilated.

Insect eggs are not appreciably affected by most fumigants, and if eggs are present it is therefore necessary to repeat the treatment after such eggs have hatched, and before the insects have become sufficiently adult to lay more. Especially in the case of White Fly, adult, immature stages and eggs may be found together on the plants, and it is often necessary to fumigate three times, at intervals of about ten days, in order to eradicate this insect.

APPENDIX I

OTHER ARTHROPODA OF ECONOMIC IMPORTANCE

CLASS I. CRUSTACEA

THE vast majority of Crustaceans are aquatic, and like the crabs, lobsters, shrimps, etc., have appendages which show great variation in different segments of the body.

One order, the ISOPODA, consists of animals much flattened dorso-ventrally, and possessing walking appendages which are all very much alike. Included in this order are terrestrial forms belonging to the family *Oniscidæ*, which go by the name of Wood-lice, Slaters, Sow Bugs or Pill Bugs.

Adult **Woodlice** are slaty grey in colour, with usually seven pairs of legs, and antennæ of seven or eight joints. The female produces eggs early in summer, but retains them in a special pouch, called the *marsupium*, formed by the overlapping of certain chitinized plates on the anterior ventral surface of the body. Incubation takes from fifty-six to ninety-three days according to species, and there are from thirty to eighty individuals per brood.

The young slaters are at first pure white in colour, with only six pairs of legs, but after the second moult, which occurs fourteen to eighteen days after hatching, the seventh pair of legs appears. The number of moults depends upon the food supply.

Woodlice live in damp situations only, and usually avoid sun-light as much as possible. They feed upon vegetable matter, and may cause serious damage to plants in glasshouses and potting-sheds, whilst some species attack strawberries and wall fruit. There are thirty-five species in the British Isles, only about ten being of economic importance.

The two commonest species are *Oniscus ascellus* and *Porcellio scaber*. The first may be as long as 16 mm., and has antennæ of eight joints. It is found chiefly in glasshouses, but it has also been reported damaging strawberry roots. *Porcellio* occasionally attacks ripening wall fruit, such as peaches and nectarines: it frequently causes considerable damage in cold frames belonging to Channel Island growers. It has seven-jointed antennæ.

Another common species is *Armadillium vulgare* (Fig. 135). This rolls itself into a ball at the least provocation, whence is derived its name ' pill-louse.' It has been found at work upon potatoes, beans and mushrooms, and there is reason to believe that this species frequently causes damage for which slugs are held to be responsible. It often eats off tomato seedlings at soil level.

Control Measures.—Woodlice can be killed in large numbers by placing here and there sliced potatoes smeared with Paris green, or the poison can be put direct upon the floor and covered with damp boards. Dusting round garden beds with a mixture of equal parts of Paris green and ground lime has been recommended. The loose strawy material which accumulates round rubbish heaps forms breeding grounds for the animals, and could be burnt with advantage. Woodlice in glass-houses may be destroyed by hot water, or by a spray consisting of 2 pints of the following mixture in 12 gallons of water: Cresylic acid, 1 gallon; soft soap, 8 lbs.; naphthalene, ½ lb. This should be forced into all cracks and crevices in the greenhouse.

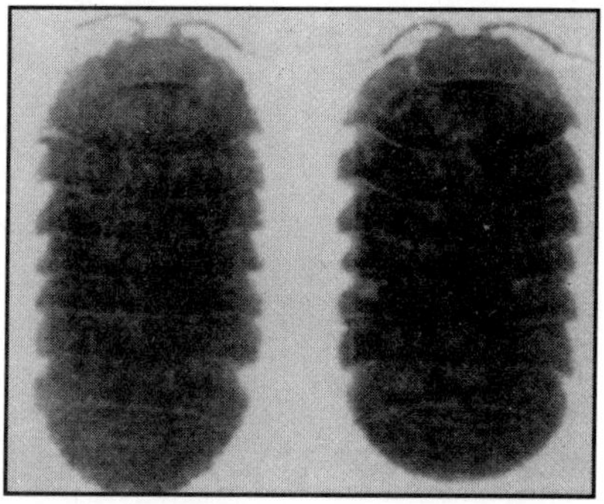

FIG. 135.—Common Woodlouse (*Armadillium vulgare*). Reproduced by permission of H.M. Stationery Office and the Ministry of Agriculture and Fisheries.

CLASS II. ARACHNIDA—SPIDERS, MITES, ETC.

The class Arachnida includes the Scorpions, Spiders, False Spiders, Harvestmen, Mites and Ticks. Spiders (*Araneæ*) may be distinguished from Mites (*Acari*) by the following peculiarities:

Araneæ.—Body divided into two segments, namely *cephalothorax* and abdomen, the anterior part of the body consisting of the head and thorax fused together (Fig. 136, II).

Acari.—Body unsegmented, possessing eight legs in the adult stage and six in the larval condition: mostly of very small size (Fig. 136, I).

Spiders have a certain economic importance, since they destroy numbers of insects which are harmful to or annoy man. The best-known species construct webs, but others hunt their prey. It should be noted that the so-called 'red spiders' do not belong to the Araneæ, but are true mites.

ACARI OR MITES

The Acari or Mites are small or microscopic arthropod animals, frequently less than 1/200th of an inch in length. They have no spiracles, but breathe through the whole of the body surface: eyes are absent. The young mite which hatches out from the egg possesses three pairs of legs and is called a *larva*. After a number of moults the fourth pair of legs appears, but the animal is still without developed sex organs, and is known as a *nymph*. After further moults the nymph becomes sexually mature. The

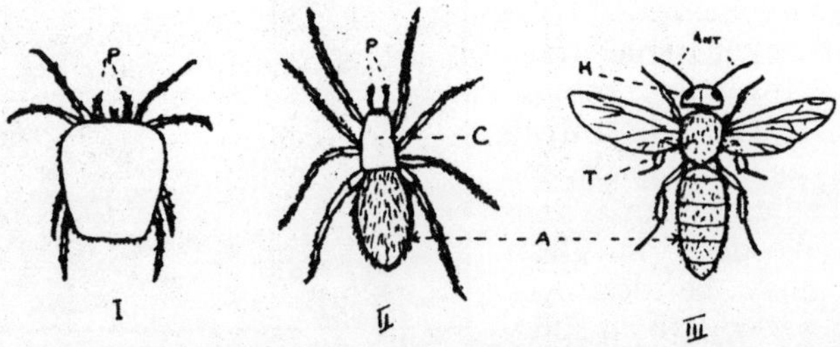

FIG. 136.—Diagram showing differences between (i) Mite, (ii) Spider, (iii) Insect. *A*. abdomen: *Ant*. antennæ: *C*. cephalothorax: *H*. head: *P*. palps: *T*. thorax.

Gall Mites, however, have two pairs of legs only. Some mites reproduce viviparously. Mouthparts are constructed for biting and piercing.

Mites must be numbered amongst the most destructive and harmful of animals. Not only do they attack both domestic animals and plants, but they serve as carriers of some of the most dreaded diseases of man and farm stock. Some species burrow in the skin of their hosts, causing intense irritation and the formation of scabs; others suck blood, and some live in the breathing tubes of insects. A number of species attack domestic fowls, living on their legs or amongst their feathers.

From an agricultural and horticultural point of view the most important families of Acari are the following:—

1. *Eriophyidæ*, or Gall Mites.
2. *Trombidiidæ*, or 'Red Spiders.'
3. *Ixodidæ*, or Ticks.
4. *Gamasidæ*, including the Red Fowl Mite.
5. *Sarcoptidæ*, or Itch Mites.

6. *Demodecidæ,* the cause of follicular mange.
7. *Tarsonemidæ,* the cause of acarine disease of hive bees.
8. *Tyroglyphidæ,* or Cheese Mites.

1. FAM. ERIOPHYIDÆ, or **Gall Mites.**—These are distinguished from the other families on account of their tapering bodies, and the presence of only four legs. The two posterior pairs of legs are wanting. The skin is much wrinkled and a number of long hairs are sparsely situated on the body (Fig. 137). They live in buds or upon the leaves of plants.

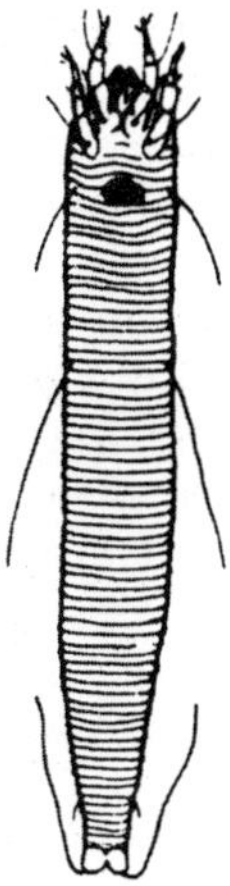

FIG. 137.—Currant Gall Mite (*Eriophyes ribis*).

The Black Currant Mite (*Eriophyes ribis*) is the most important species. It is the cause of the disease in black currants known as 'big bud.' This complaint is characterized by the swelling up of attacked buds until they are double the normal size. They are very noticeable in spring, and they usually fail to produce leaves and flowers. Examination of diseased buds reveals the presence of large numbers of gall mites, each specimen being about one-hundredth of an inch long. The mites enter the buds during July, and remain in them until the following March, during which period they lay large numbers of eggs. From March onwards, owing to the drying up of the distorted buds, the mites migrate over the twigs, and may get carried to other bushes by wind and other agencies. They live at the base of leaf stalks until the developing buds are large enough to enter.

Red currants and gooseberries are occasionally attacked, but the nature of the buds of these plants does not permit such easy entry as those of the black currant. It is the young growing shoots of gooseberries and red currants which are principally affected: they become swollen to double their normal circumference.

Control Measures.—Diseased stock should never be planted, nor should cuttings be taken from diseased bushes. Badly infested bushes are best grubbed up and destroyed by fire. If one or two bushes only are slightly affected, the disease may be kept in check by pruning off the diseased shoots early in the year and burning them. Dusting the bushes with a mixture of lime and sulphur has given good results in some cases. At the end of March the first dusting, consisting of equal parts of slaked lime and sulphur, is given, whilst the bushes are damp, and three more applications are given at fortnightly intervals. This kills or hinders the mites during their migratory period. Spraying with winter-strength lime sulphur when the young leaves are about the size of a shilling has been found beneficial. Slight scorching of the leaves may result from this rather late spraying, but no permanent injury should result if the operation is carefully performed.

2. FAMS. TROMBIDIIDÆ and TETRANYCHIDÆ.—**Red Spiders and Spinning Mites.**—These two families are closely allied, and include a number of species destructive to fruit bushes, hops, and greenhouse and hothouse plants. Members of the Tetranychidæ spin small webs.

The Gooseberry Red Spider (*Bryobia ribis*) attacks the leaves of gooseberry bushes, piercing the epidermis and sucking the sap. The leaves become a characteristic marbled grey colour, or almost silvery; the young ones remain stunted, and frequently leaves, flowers and even the young fruit drop off. The adult mite is about one-thirtieth of an inch long, and its first pair of legs is very much longer than any of the others. In colour it varies from red to a greyish green. The females lay globular, reddish eggs on the twigs in March and April. These hatch out in four or five days, and in a fortnight the animals become mature. They are most active during bright weather. Hibernation takes place both in the egg and the adult forms. The ivy web-spinning spider (*B. prætiosa*) is often confused with this species.

Nicotine, liver of sulphur, or paraffin emulsion sprays may be successfully employed against this pest.

The Red Spider of the vine and the Hop Red Spider belong to the Tetranychidæ. They may be distinguished from the species described above by the fact that their legs are all of the same length, and shorter than those of the gooseberry red spider. Both these species spin silken webs on the underside of the leaves of the plant attacked, and in these webs eggs are laid and the young mites develop.

Under glass red spiders are greatly encouraged by over-heating and excessively dry conditions. Fumigation with sulphur or tobacco destroys them.

Included in the Trombidiidæ is the **Harvest Bug**, which is the larval or six-legged form of a Trombid mite. It fixes itself at the base of a hair of men and animals, generally on the limbs, and causes a most painful irritation as a consequence of its bites. Bathing the affected parts with ammonia gives relief from the itching. In America these Harvest Bugs are called 'chiggers': they attack domestic fowls, and may transfer their attentions to the men who handle them.

3. FAM. IXODIDÆ.—**Ticks** are large mites provided at the anterior end with a special piercing organ or rostrum, bearing a number of teeth or barbs which point backwards. Onee a tick has thrust its rostrum into an animal, it is impossible to pull it away without leaving the head behind. Ticks breathe through spiracles. Their bodies are leathery, and that of the female is frequently capable of enormous distention. The female is usually larger than the male. Ticks may cause damage to man and farm stock in three ways: 1. By causing irritation through their bites; 2. By sucking blood; 3. By transmitting disease. It is in their capacity as disease carriers that ticks are most destructive. Amongst the more important tick-borne diseases are the African tick fever in human beings, malignant jaundice of the dog, heartwater in sheep and goats, and redwater in cattle. In this country ticks are not strongly represented, but they causc great losses in Africa, Australia and America.

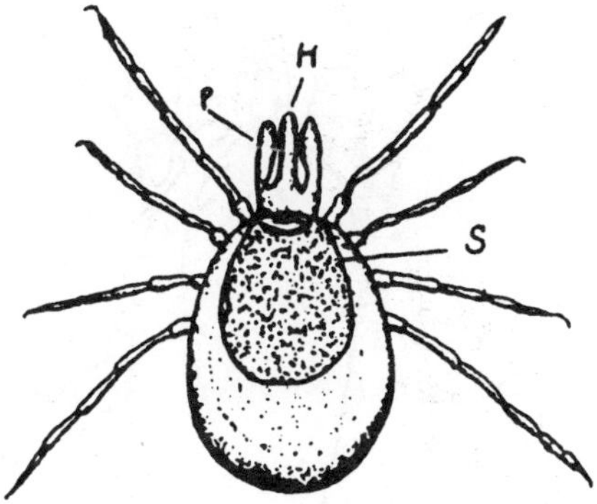

FIG. 138.—*Ixodes ricinus* (female). S. scutum: H. hypostome: P. palp.

There are two kinds of ticks.

(a) *Ixodidæ*, or Hard Ticks.—These have the whole of the dorsal surface in the male covered with a chitinous shield or *scutum*; in the female the scutum only partially covers the body. The rostrum is at the anterior end of the body and visible from above. *Ixodes ricinus* (Fig. 138) attacks sheep in the Border districts and in the Western Highlands. It is killed during dipping.

(b) *Argasidæ*, or Soft Ticks.—These have no dorsal shield, and the rostrum is ventral and invisible from above. A number of these soft ticks attack birds, the pigeon and fowl in particular. In America, Persia, India and elsewhere a tick (*Argas persicus*, = *A. miniatus*) attacks fowls, and spreads the blood disease known as *spirochætosis*.

4. FAM. GAMASIDÆ.—These mites are closely related to the ticks, but are very much smaller in size and have their bodies covered with chitinous plates. They are frequently met with on rodents and upon birds.

The Red Fowl Mite (*Dermanyssus gallinæ = D. avium*) is the most important representative in this country. This is about ·5 mm. in length, and is provided with very long piercing mouth-parts (Fig. 139). In colour it varies from dirty white to red, according to the amount of blood it has recently sucked from its host. The female lays its eggs in crevices in fowl-houses or nesting-boxes, or in filth of all kinds. The larvæ are white, and the skins discarded whilst moulting may frequently be seen as a white powder on perches, etc. Fowl mites are nocturnal feeders, hiding during the day-time and migrating on to the birds at night. They cause the birds to lose condition through lack of rest and loss of blood. They occasionally get on to the hands of persons handling fowls and may occasion severe itching for a time. Horses may also be worried by these pests. They are seldom met with in well-ventilated, clean houses: the measures recommended for getting rid of lice and fleas (p. 56) may be adopted against the fowl mite.

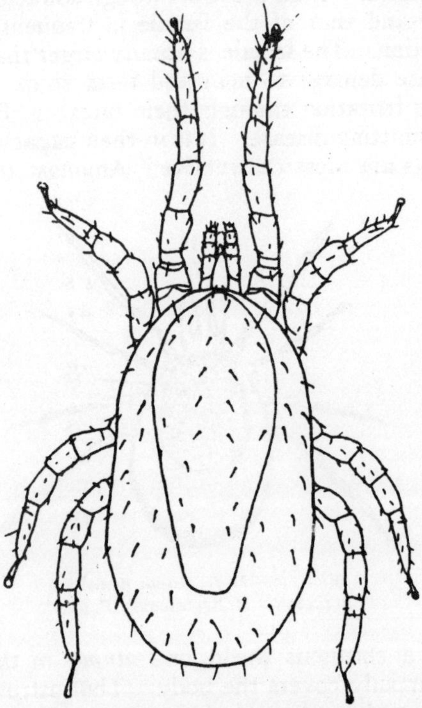

FIG. 139.—Red Fowl Mite (*Dermanyssus gallinæ*). Copied from British Museum Guide.

5. FAM. SARCOPTIDÆ.—**The Itch Mites** are the smallest Acari known, ranging from 1/250th to 1/25th of an inch in length. They attack man and animals, living in or on the skin, causing crusts or scabs to form. The diseases caused by the itch mites are known as 'scabies' and parasitic mange. All the Sarcoptidæ lay eggs, but some are viviparous. The females are much larger than the males. The fourth pair of legs is always much shorter in the male than in the female. There are three important genera belonging to this family, distinguished as follows :

1. *Psoroptes.*—Body round: terminal sucker of the legs borne on a long, three-segmented peduncle (Fig. 140, I). These mites do not burrow into the skin, but live under the scabs which they form.
2. *Sarcoptes.*—Body round: terminal sucker on a long, simple peduncle; posterior legs almost or quite concealed beneath the body (Fig. 140, III). These mites burrow into the skin, beneath the surface of which the females deposit their eggs.
3. *Chorioptes* (formerly called *Symbiotes*).—Body oval: terminal sucker on a short, simple peduncle (Fig. 140, II). Chorioptic mites live under scabs after the style of sarcoptic species, but they are much more localized and do not spread all over their host.

Sheep in particular are susceptible to the attacks of itch or scab mites, and the life histories and habits of these species may be taken as typical of the family.

Common Scab of Sheep.—This disease is caused by *Psoroptes communis*, var. *ovis*, a mite about one-fortieth of an inch long (Fig. 140, I). The male is smaller than the female and has suckers on all legs, the third pair of legs being very short; the female has long hairs on the third pair of legs in place of suckers. The females lay a number of eggs which hatch in about seven days. The young mites commence to feed upon the outer layers of the skin, setting up violent irritation, which leads to the exudation of a yellowish discharge and the formation of thick, greasy crusts or scabs. The sheep fall off in condition and the wool frequently falls out in large patches. Psoroptic scab occurs chiefly on the woolly regions of the body, and is seldom found on the extremities. The mites, however, may remain in the ears of sheep for some time before spreading over the body. The disease is highly contagious, and is spread by sheep rubbing themselves against fences, etc., as well as by actual contact with infected animals.

Sarcoptic Scab of Sheep.—This mite (Fig. 140, III) is confined chiefly to the regions round the head, though in advanced cases it may spread over the body. It is less common than psoroptic mange, but is more difficult to eradicate, because the parasites burrow into the skin before the scab is formed.

Chorioptic or Symbiotic Scab.—The mite causing this form of mange (Fig. 140, II) restricts itself to the extremities of the sheep. It is found chiefly on the feet and round the base of the tail: it spreads very slowly, and is of a mild type.

Control Measures.—The disease comes under the regulations of

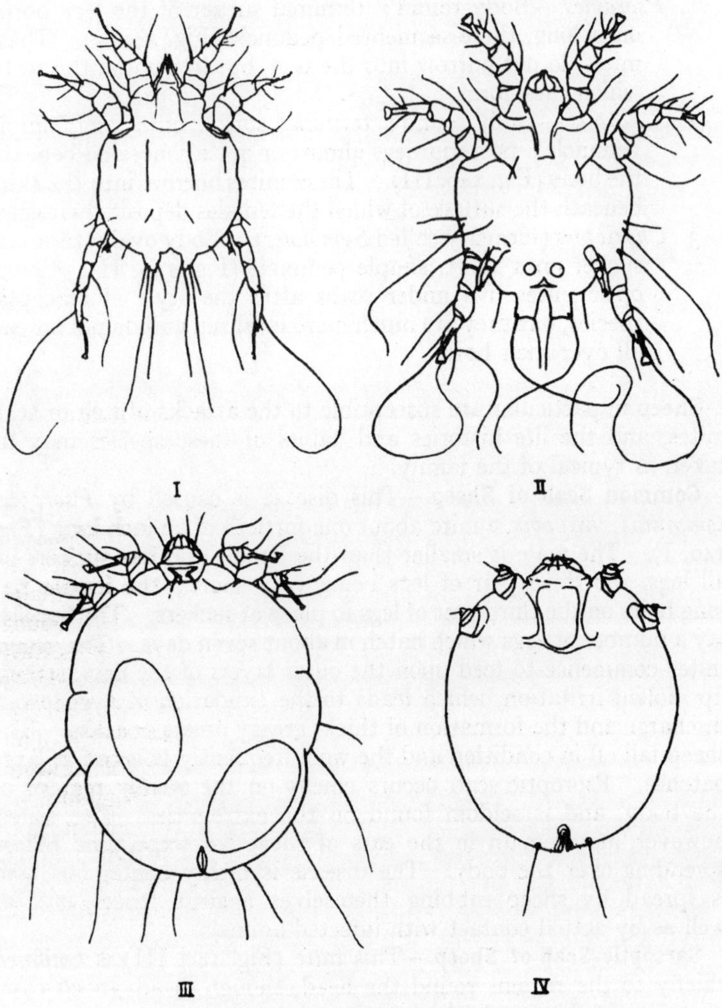

FIG. 140.—Scab Mites (much enlarged but not all drawn to the same scale).
I. *Psoroptes communis*. II. *Chorioptes ovis*. III. *Sarcoptes scabei*.
IV. *Cnemidocoptes mutans*.
Redrawn from British Museum Guide.

the Sheep Scab Order of 1920, and is notifiable to the competent authorities whenever discovered. Dipping is universally practised to kill the casual parasites, and to be efficient a second dipping must be given not sooner than seven days, and not later than ten days, after the first. This ensures the death of those mites which hatch out in the interval from the eggs situated beneath the scabs, because the dip solutions do not kill eggs. Many makes of dips are on the market and have been passed as efficient by the Ministry of Agriculture and Fisheries. The chief constituents of these proprietary articles are arsenic, sulphur, tobacco and carbolic acid. A swim bath is the most efficient way of using dipping solutions, since the sheep can be completely immersed in a safe and expeditious manner.

The horse, mule, dog and pig, as well as cattle, are subject to attack by the varieties of the three kinds of mite mentioned above. Parasitic mange in horses, asses and mules comes within the province of the Parasitic Mange Order of 1911. Strict isolation and cleanliness, combined with local applications of acaricides, are necessary to cope with the complaint.

Closely allied to the genus *Sarcoptes* is the genus *Cnemidocoptes*, in which all the legs of the male end in a sucker borne on an unsegmented peduncle, whereas in *Sarcoptes* the third pair of legs has no sucker. Two important species which attack domestic fowls require notice.

Sealy-Leg in Fowls.—*Cnemidocoptes* (*Sarcoptes*) *mutans* (Fig. 140, IV) is the cause of this common complaint. The mite attacks the feet and legs of poultry, boring under the scales and deep into the underlying tissues. The eggs hatch in about ten days. Considerable irritation is caused, and the scales become raised at the edges by the formation of a floury powder often mixed with exuded serum. In bad cases the birds go lame, and the feet may even drop off. The disease is highly contagious, and affected birds should at once be isolated. A rapid, if somewhat drastic, method of cure is to dip the legs and feet into a mixture of one part paraffin oil and two parts linseed oil each night. An ointment consisting of one part oil of caraway, and five parts white vaseline, has also been found very effective and less dangerous than the above treatment.

Depluming Scabies is caused by *Cnemidocoptes* (*Sarcoptes*) *lævis*, var. *gallinæ*, which lives at the base of the quills. It causes the feathers to break off at the surface of the skin, and also encourages the bird to pluck out its feathers in an attempt to allay the irritation. It occurs chiefly in spring, starting at the rump and spreading over the back and breast. Oil of cloves, or a mixture of one part creosote to twenty parts lard or vaseline, rubbed into the affected areas, kills the parasite.

6. FAM. DEMODECIDÆ.—Members of this family are exceed-

ingly small and wormlike, the posterior region of the body tapering into an elongated cone. The legs are borne upon the rather broader anterior region, whilst the posterior region is transversely wrinkled. These mites live in the hair follicles and sebaceous glands of many kinds of farm animals, giving rise to demodectic, follicular or red mange. The most common species is *Demodex canis*, which causes follicular mange of the dog (Fig. 141). The mites burrow deeply into the follicles and sweat glands, and are usually associated with a bacterium (*Staphylococcus pyogenes albus*). The two irritants give rise to reddish pustules and crusts, whilst a very characteristic and disgusting odour comes from the affected regions. Treatment is difficult owing to the deep-seated nature of the disease; those acaricides compounded with horse fat appear to be most successful, since horse fat is very readily absorbed by animal tissues. The horse, cow and sheep are at times attacked by species of *Demodex*, but cases are not common.

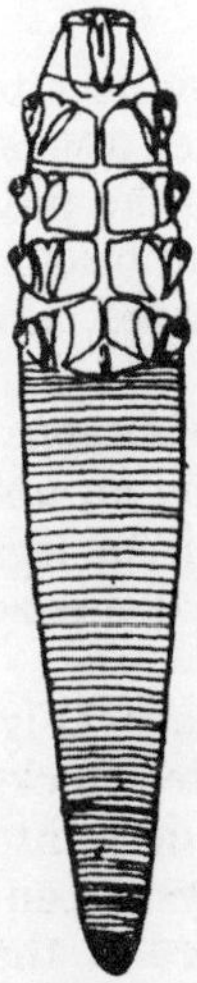

FIG. 141.—Demodectic Scab Mite (*Demodex canis*).

Redrawn from British Museum Guide.

7. FAM. TARSONEMIDÆ. — **The Isle of Wight Disease** of hive bees is caused by *Acarapis woodii*, a member of this family. The mites live and breed in the tracheal tubes of the thorax, which turn brown or black. Affected bees lose the power of flight and become 'crawlers,' many of them eventually dying. Infection takes place through the spiracles, and is spread by bees visiting the same flower and during the 'robbing' of a hive by strange bees. Weak stocks affected by the complaint should be destroyed by burning sulphur in the hive, and every effort should be made to prevent 'robbing.' Everything which has been in contact with diseased bees should be carefully sterilized.

8. FAM. TYROGLYPHIDÆ.—These are soft bodied, pale coloured mites which live in flour, grain, cheese, etc. Some species exhibit a curious form known as the *hypopus*, which occurs after the larva has moulted into the nymph stage. The hypopus resembles a tortoise in shape, and has on its ventral surface a sucker plate by means of which it attaches itself to flies and human beings. In this way it is carried from one place to another, only to resume feeding at a favourable opportunity. Four species attack cheese, particularly Stiltons. Eggs hatch in ten to twelve days, and the mite becomes mature in four to five weeks. The mites breed in

cracks in shelving, etc., and may also be carried on to the cheeses by flies: they are more abundant in dark store rooms than in light ones.

Prevention is the best method of dealing with this mite; all shelves, tables, etc., should be scrubbed with a five per cent. solution of carbolic acid between one Stilton season and the next, and the store room should be netted against flies. A daily brushing of those Stiltons affected by mites considerably lessens the damage.

Class III. MYRIAPODA

This class of animals is very closely allied to the Insects, for although the bodies of Myriapods are wormlike, their internal structure is very similar to that of the Hexapods. They breathe through tracheal tubes, possess antennæ, and never have more than one pair of legs on each of the first three segments. There are two groups of Myriapoda.

1. *Chilopoda* or **Centipedes.**—These have a rather flattened body consisting of a large number of segments each bearing one pair of legs. The mouthparts consist of a pair of biting mandibles, a pair of maxillæ, a pair of leg-like palps, and finally a pair of powerful, biting, poison jaws. The generative organs open on the last segment but one (Fig. 142).

2. *Chilognatha* or **Millipedes.**—These are composed of numerous rounded segments, each, with the exception of the first three, bearing two pairs of legs. The mouthparts closely resemble those of a mandibulate insect. The generative organs open between the second and third segments of the body (Fig. 143).

Centipedes.—These are active, carnivorous animals living upon insects, small slugs and snails. They are consequently beneficial. There are two very common genera in this country, *Lithobius* and *Geophilus*. *Lithobius* is large and stout, brown in colour. The female lays eggs, and whilst they are still moist, rolls them in soil so as to conceal them. It is stated that this is necessary because the male at once devours any eggs that he sees. *Geophilus* is longer and narrower, about 1½ to 2 inches long: it is the commonest centipede met with in the soil. *G. electricus* is phosphorescent, yielding quite a bright light.

Millipedes.—These are vegetable feeders, and frequently do considerable damage by eating roots, bulbs, tubers, etc. They cause a false 'scab' in potatoes. There are two common genera—(1) *Julus*, of which *Julus* (*Blanjulus*) *pulchellus* is the commonest

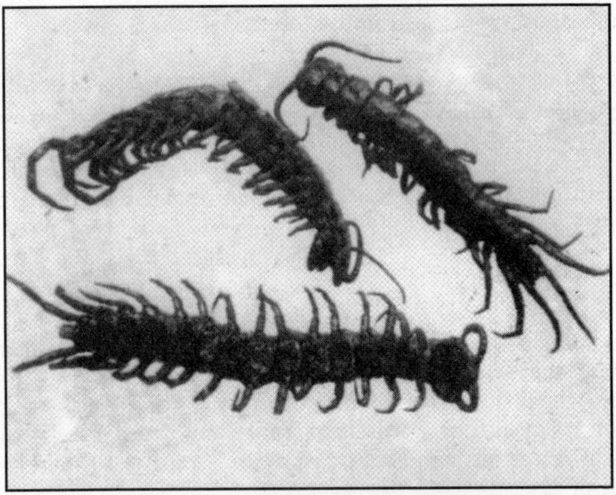

FIG. 142.—Centipedes (*Lithobius* spp.).

FIG. 143.—Millipedes (*Julus terrestris*).

species, consisting of thirty or more smooth cylindrical segments; (2) *Polydesmus*, which has about twenty, broader, flatter segments. *J. pulchellus* is about half an inch long, yellowish, with crimson spots along either side. *J. terrestris* is a larger species, of a leaden black colour, with whitish legs.

Polydesmus complanatus is nearly an inch long, varying from a pale purplish white to a dull red or brown colour.

Millipedes breed from May to July, laying from sixty to one hundred eggs in a spherical nest, made by gluing together particles of earth with a kind of saliva. Hatching occurs in about a fortnight. Young millipedes have at first only six legs, situated on the anterior segments. Growth takes place at the hinder end, new segments being added between the last segment and the one before it. Millipedes are slower in their movements than centipedes, and may roll themselves up into a spiral on being disturbed.

Control Measures.—If the infestation is not too severe, control may be achieved through trapping with poisoned baits, such as pieces of potato and mangold soaked in Paris green.

Naphthalene as a soil fumigant has also been successfully used. Millipedes sometimes breed in enormous numbers in garden manure heaps: such manure should not be applied to the garden until it has first been treated with naphthalene or carbon disulphide to kill the pests. Poultry will clear off large numbers of millipedes if they get the opportunity.

APPENDIX II

ANGUILLULIDÆ OR EELWORMS

EELWORMS are in no way related to insects: they belong to an entirely different group of animals, namely the phylum Vermes or Worms. They are included here because some of them attack plants and cause diseases resembling those brought about by the activities of insects; a knowledge of the structure and habits of eelworms is consequently essential to the agricultural entomologist.

The phylum Vermes includes both flat worms and round worms. Eelworms form the *Family* Anguillulidæ of the *Class* Nematoda, which embraces certain types of round, unsegmented worms. The structure of a typical nematode is shown diagrammatically in Fig. 144. The body is long, narrow, and cylindrical, comparatively blunt at the anterior end and pointed posteriorly. It is covered by a more or less transparent cuticle and muscular layer.

The alimentary canal runs ,without distinct convolutions from mouth to anus. The mouth is surrounded by three lips, and opens into the œsophagus. This may, or may not, possess a distinct muscular swelling in the middle of its length: sometimes the posterior end of the œsophagus swells out into an enlarged glandular portion. There is a nerve ring round the œsophagus, from which a number of nerves run both anteriorly and posteriorly.

The rest of the gut consists of a tube, somewhat larger than the œsophagus, which finally opens via the anus, situated towards the hinder extremity.

In eelworms the pore to the excretory system is situated roughly on a level with the hind end of the œsophagus. There are no vascular and respiratory systems.

The sexes are distinct, and *sexual dimorphism* is common: that is, the male often differs markedly in appearance from the female. The *gonad* (testis or ovary) is tubular and lies alongside and ventral to the alimentary canal. The ejaculatory duct of the male opens close to the anus, but the vulva of the female is usually anteriorly situated.

The males of certain species of eelworm possess curious accessory sexual organs. At the posterior end the cuticle expands into small wing-like projections known as *cuticular alæ*,[1] giving this region a bluntish appearance (Fig. 146). One or more pointed, slightly curved *copulatory spines* or *spicules* may also be present.

Eelworms lay comparatively large eggs, and the sexually immature forms which hatch out are called larvæ.

Eelworms are minute animals, rarely exceeding one-fifth of an inch in length. By far the greatest number live freely in water or in damp soil, whilst comparatively few species parasitise cultivated and wild plants. Moisture indeed seems to be the factor which determines the prolificacy or otherwise of eelworms, although the endoparasitic forms are not so susceptible to dryness as the free-living species. Under dry conditions the larvæ of eelworms are capable of encystment, and in this cystic state they can successfully endure considerable desiccation. Eelworms are most numerous in the upper layers of moist, shady soil: dry sunny soils contain fewer numbers. If buried too deeply, as, for example, by extra deep ploughing, large numbers of eelworms perish. They do not appear to be very adversely affected by frost unless it is prolonged, and they are consequently able to survive most winters. The largest numbers of sexually mature forms are to be found usually in spring and autumn.

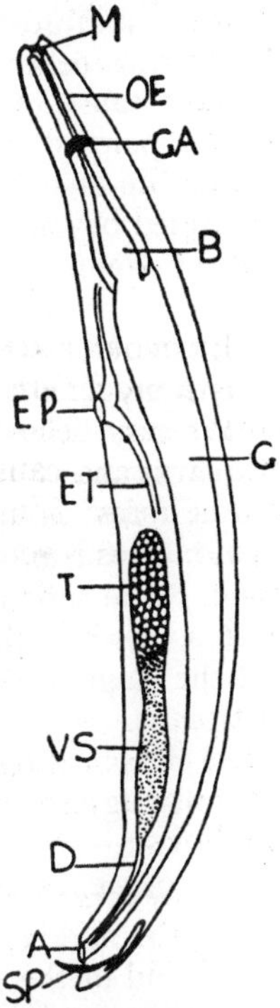

FIG. 144.—Diagram of the structure of a male nematode.

M. mouth : *Œ*. œsophagus : *GA*. nerve ring : *B*. bulb : *Sp*. spine with sheath : *A*. anus : *D*. ejaculatory duct : *VS*. seminal vesicle : *T*. testis : *ET*. excretory tube cut short : *EP*. Excretory pore : *G*. gut. (Thomson.)

Those eelworms which are found in the roots, stems and leaves of cultivated plants cannot all be strictly described as parasites. It has been proposed to classify eelworms roughly into three groups, namely, *free living*, *semi-parasitic* and *parasitic* forms. *Free-living* forms are not usually found in plants.

[1] This expanded region has been called by Marcinowski and others the *bursa*. The true bursa, however, as found in the Strongylidæ, consists of cuticular expansions supported by a definite system of 'rays.' In cuticular alæ these 'rays' are absent.

Semi-parasitic forms are frequently found in large numbers in plants, but they gain their entry usually through existing wounds or diseased areas; they are not usually capable of penetrating healthy, uninjured plant tissue.

Parasitic eelworms can penetrate plant tissues, and are the primary cause of decay in attacked plants. They all possess a curious pointed spine (*stylet* or *buccal spear*) which normally lies within the mouth cavity, but which some observers state can be protruded beyond the lips to assist in perforating the epidermis of root or stem. Other observers deny that this movement can occur.

The mere presence of eelworms in a diseased plant is not necessarily a proof that these animals are the cause of the disease. The worms may be semi-parasites, or there may be one or two parasitic forms and hundreds of semi-parasites. Parasitic worms may also be found in apparently healthy plants. The conditions most favourable for infection by parasitic eelworms are not fully understood. The seriousness of eelworm attack depends upon several factors, of which the most important are the numbers of eelworms, and the stage of development and healthiness or otherwise of the host plant.

The classification of Anguillulidæ is still a matter of dispute. It is generally agreed, however, that there are three parasitic genera, as follows:—

1. *Tylenchus*,[1] characterized by a muscular bulb or swelling in the middle, and a swollen glandular region at the posterior end of the œsophagus (Fig. 145, *A*). The tail of the male possesses narrow cuticular alæ. The buccal spine is knobbed at the base. When killed by gentle heat the male stretches out quite straight.
2. *Aphelenchus* has no clearly defined second portion to the œsophagus (Fig. 145, *B*). The tail of the male has no cuticular alæ, and the buccal spine is without basal swellings. When killed by gentle heat the tail of the male curls up towards the ventral surface.
3. *Heterodera*.—In this genus the buccal spear is massive, and the females are much swollen and citron-shaped (Fig. 147).

(*N.B.*—Both *Tylenchus* and *Aphelenchus* contain many more free-living than parasitic species.)

The most important species, belonging to the above genera, which attack cultivated plants are described below:

[1] Also known as Anguillulina.

GENUS I.—*Tylenchus.*

The Stem Eelworm (*Tylenchus dipsaci* = *T. devastatrix*).

The worm is shown in Fig. 146. It is from 1.2 to 1.5 mm. long, and its buccal spine is from 12 to 14 μ long. It attacks a large variety of cultivated plants, cereals, bulbs, onions, clovers, and occasionally potatoes. In all cases it is the stem of the host which is attacked, the roots not being penetrated.

The worm on emerging from the egg has to undergo a series of moults before becoming sexually mature. The rapidity with which

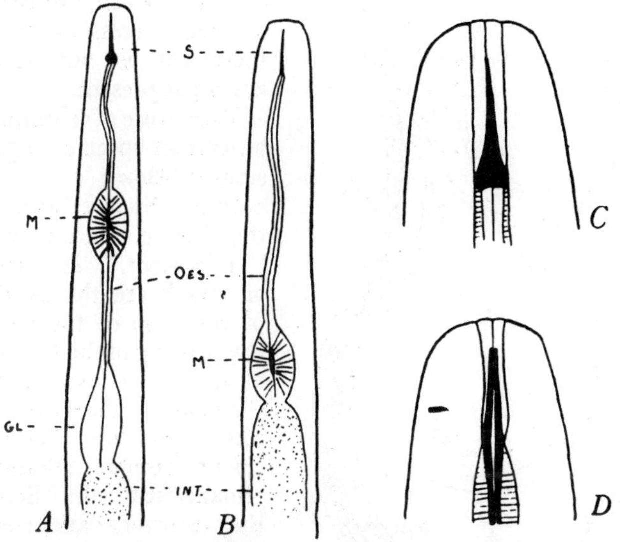

FIG. 145.—Anterior regions of Eelworms. *A.* of *Tylenchus dipsaci.* *B.* of *Aphelenchus fragariæ.* *C.* Buccal spine of *T. dipsaci.* *D.* of *A. fragariæ.*

S. buccal spine: *M.* œsophageal bulb or swelling: *Œs.* œsophagus: *Gl.* glandular portion of œsophagus: *Int.* intestine.

After Marcinowski.

these moults occur depends very largely upon the rate at which the host plants are growing. If plant growth is rapid, the parasite grows correspondingly quickly. Temperature appears to have little direct effect upon the eelworm except in so far as it affects plant growth.

The effect of increasing numbers of eelworms is to cause hypertrophy of plant tissue, with much crinkling of leaves, distortion, chlorosis, and splitting of the foliage. In many cases death of the host plant is the result, and the tissues dry out.

Under these dry conditions the eelworms become quiescent: they can remain in this desiccated state for very long periods, becoming active again when moistened.

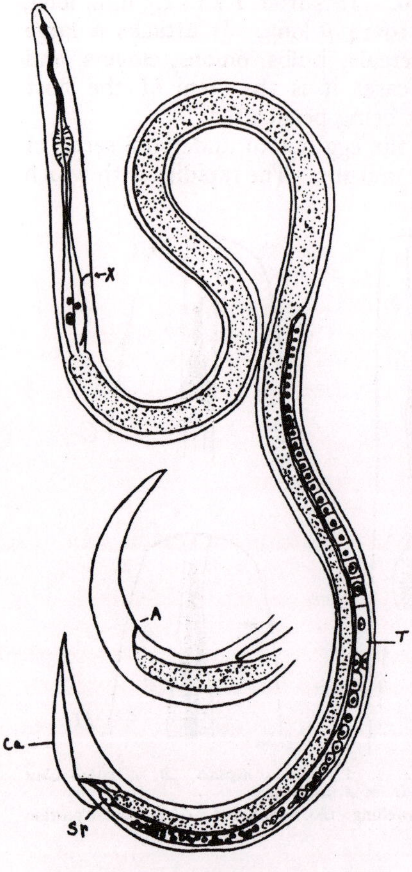

FIG. 146.—Stem Eelworm (*T. dipsaci*). Male, showing internal anatomy and cuticular alæ. Inset, tail of female. *A*. anus. *Ca*. One of the cuticular alæ: *T*. testis: *X*. excretory pore: *Sp*. spicule. After Marcinowski (modified).

Those worms which enter the soil before the drying out of the host plant are not capable of living indefinitely without a further supply of food. If no suitable host plants are available they perish, but the length of time necessary to starve them out is not accurately known at present.

The nature of eelworm attack on certain specific crops is described below:

Oats.—Stem eelworm causes the disease in oats known as 'Tulip Root.' The symptoms of attack are the swelling up of the base of the stem, and the twisting of the leaves, which also become wrinkled and corrugated along the edges. The plants lose their healthy colour, become chlorotic, and remain stunted. Sometimes they turn very dark green in the early stages of attack. Many of them fail to produce ears, and the whole crop may fail if the infestation is severe. The disease is very prevalent along the north-east coast of Scotland.

Infection takes place from the soil. Robertson also states that the worms may frequently be found between the caryopsis and pales of oat 'seed'; he recommends therefore that seed from an infected crop be burnt. It certainly should not be used for seed, and there is a danger that if used for feeding purposes the worms may pass unharmed through the alimentary canal, and get back to the field in manure.

Oats may be hastened through a slight attack by top dressings of nitrogenous and potassic fertilizers. Extra deep ploughing after an attack might assist in reducing the numbers of active eelworm in the soil.

Clover.—One form of clover ' sickness ' is caused by *T. dipsaci.* Attacked plants remain stunted and deformed, whilst the stipules and bases of the leaf-stalks exhibit a characteristic swelling. The new leaves have very short stalks, whilst the blades are twisted, wrinkled and discoloured. The disease is most apparent during the winter months: eventually the clover dies out in patches.

Red clover and alsike clover are the species most frequently attacked. White clover is much less susceptible. The disease is best prevented by not growing either of these plants too frequently on the same land. To grow red clover on a field more often than once in every six or eight years is to invite ' sickness ' either from eelworm, or the fungus *Sclerotinia.* Infected crops may be stimulated by nitrogenous and potassic manures.

Bulbs.—Onions are sometimes attacked by stem eelworm, with serious results. The tissues become swollen and distorted, and rot sets in when such diseased bulbs are stored.

Narcissus bulbs also suffer severely from eelworm attack. Infected bulbs start growth abnormally early, and generally have a number of crinkly leaves growing at right angles to the bulb. The worms travel up and down the leaves, which become discoloured and finally drop, whilst the flower either does not appear or loses much of its beauty. The disease appears to be spreading rapidly owing to the ease with which the worms may be disseminated in unhealthy bulbs.

The parasites may be starved out of an infested soil by removing all bulbs from it for several years. This is a slow process and it is better to prevent the disease by planting only worm-free bulbs. Any bulbs coming from suspected areas should first be subjected to the hot water treatment, which kills any eelworms in the bulb. Well ripened, dormant bulbs are immersed for three hours in water kept at a temperature of 110° F. They are subsequently removed and allowed to dry and cool gradually. The bulbs should not be exposed to sun or strong winds during the drying process or they may be considerably damaged.

Experimenters have found considerable difficulty in getting stem eelworms from one plant to infect plants of a different species or genus. For example, red clover has frequently been successfully grown from seeds sown in a crop of oats almost ruined by eelworm. Strawberries have been grown with success upon land heavily infested with bulb eelworm. From these and other observations the theory of different biological strains has arisen. In this it is assumed that certain strains of *T. dipsaci* have become so accustomed to one host plant that they are unable to accommodate themselves with ease to any other host. If this can be definitely proved it will obviously be of very great economic importance, since it will then be possible to starve out a strain of eelworm by a correctly planned rotation.

The Wheat Eelworm (*Tylenchus scandens* = *T. tritici*) is the cause of the disease in wheat known as 'Ear Cockles,' or 'Peppercorns.' Wheat plants attacked by this worm have many of the grains in the ear transformed into swollen, black or purplish galls somewhat like those occasioned by the bunt fungus. The worm is about one-thirtieth of an inch long. It makes its way between the haulm and sheathing-leaves below the ground. Thence it gets to the growing point of the plant, and eventually enters the immature ovary of the wheat flower. From four to twenty of these adults may enter each ovary, and the female there lays from 600 to 1500 eggs. The developing larvæ feed on the tissues of the ovary and so distort it into the 'cockle.' On cutting open a 'cockle,' hundreds of these immature worms may be seen: they retain their vitality for long periods.

Wheat eelworm has recently been very prevalent in south-west England.

The disease is spread in seed wheat. The Official Seed-testing Station reports that every season a percentage of the seed wheat examined there contains 'cockles.' On an average 3 per cent. of the samples examined every year at Cambridge are contaminated. Consequently only wheat free from 'cockles' should be used for seed purposes. Seed wheat may be largely freed from infected grain by steeping it in a large vat of water. After stirring, the 'cockled' grains float to the surface and may be skimmed off. If a 1 per cent. solution of copper sulphate is used instead of water, and the wheat is left in for three-quarters of an hour, stirring at five-minute intervals, the cockles can be removed, and the grain will also be pickled against bunt.

Genus 2.—*Aphelenchus*.

There are thirty known species of this genus of eelworm, living in the roots of plants, in moist humus, or in water. Four or five species are definitely pathogenic, causing damage to a number of horticultural plants such as strawberries, chrysanthemums, ferns, begonias, black currants, and violets.

The most important species is

The Strawberry Eelworm (*Aphelenchus fragariæ*).—The male has a maximum length of about .7 mm. and the female .89 mm. They live endo-parasitically in the stems, and ecto-parasitically in the buds of cultivated strawberries, giving rise to the disease known variously as 'Cauliflower disease,' 'red-plant,' 'red-leg' and 'small leaf.' The disease symptoms are the deep, blood-red colour of the young petioles and undersides of the unfolding leaflets in

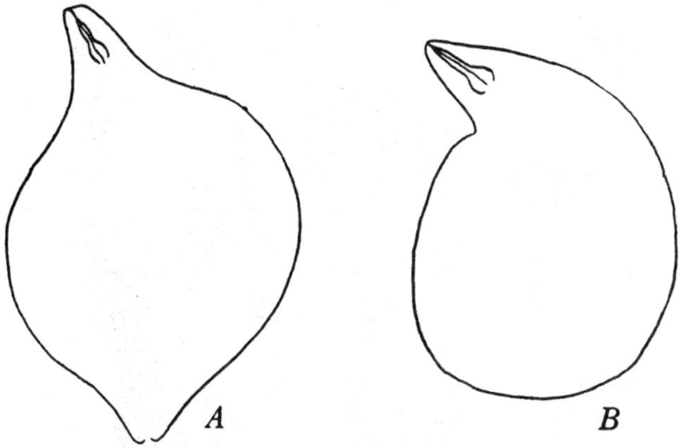

FIG. 147.—Females of *A*, *Heterodera schachtii*. *B*, *H. radicicola*. After Marcinowski.

spring, together with the swelling of the base of the petiole. The petioles may be dwarfed and the leaflets much reduced, whilst in the case of typical cauliflower disease the flowers are contorted into a cauliflower-like bunch. The crop is naturally much reduced.

So far no varieties of strawberry have been shown to be immune to the disease. On a small scale, plants may be freed from the parasites by immersing them for five minutes in water kept at a temperature of 50° C. Diseased plants should at once be removed and burnt, and in small plantations paring and burning of the top layers of soil would have a beneficial effect.

GENUS 3.—*Heterodera*.

Beet Eelworm (*Heterodera schachtii*).—This worm has for a long time caused much loss to sugar-beet growers on the Continent, and with the spread of beet culture in this country it is only to be expected that the pest may become a serious nuisance here if suitable precautions are neglected. Potatoes in certain parts of England, notably in Lincolnshire, are attacked by the same species of worm, the disease being known locally as potato 'dab.'

The male is long and narrow, not unlike the stem eelworm in shape. The female worm is shaped like a lemon or young flat fish

FIG. 148.—Roots distorted by *H. radicicola.*

('dab') and is about one twenty-fifth of an inch long (Fig. 147). She remains fixed on to the rootlets of beet or potatoes, and when mature and fertilized contains as many as 400 eggs. These usually hatch out in her body to form eel-like larvæ, but the eggs, stuck together in a jelly-like mass, are also found in the soil. The larvæ escape by the rupture of the body of the female, and proceed to bore into young roots of the beet. Those larvæ destined to become females remain permanently attached to the root they attack, but the male moves about freely in the soil. The period of development from egg to adult takes about five weeks and there may be six or seven successive generations. The larvæ retain their vitality in the soil for five or six years.

During the feeding of the larvæ and adults the outer skin of the beet is caused to bulge out into galls, and about July the leaves begin to become limp and die off. The top of the beet blackens, and decay sets in.

Control Measures.—Correet rotations, where beet or potatoes do not occupy land too often, is a great safeguard against this worm. Compost made from beet refuse known to be infected should not be used as manure, and every care should be taken not to transfer infected material from one field to another. In the case of potatoes it is the first and second early varieties which are most severely attacked, whilst the late varieties crop well in spite of the 'cysts.' Lime in conjunction with potassic fertilizers has had beneficial results.

KNOT-ROOT DISEASE of Cucumbers and Tomatoes is caused by a very closely allied species, *Heterodera radicicola* (Fig. 147), which lives deep in the roots of these plants, causing swellings and knot-like growths (Fig. 148). Infected soil in glass-houses has either to be sterilized by steam in order to kill the pest, or has to be removed. It is said that sulphate of potash kills and checks these eelworms, and that potassium permanganate (one part in 200 of water) will kill all forms but the eggs. The soil requires to be saturated every ten days for a considerable period. In glass-houses carbon disulphide may also be employed for soil sterilization.

APPENDIX III

MOLLUSCA—SNAILS AND SLUGS

THIS phylum includes the Oysters, Mussels, Cockles, and Cuttlefish as well as the common Snails and Slugs. Only one class of molluscs is of agricultural importance, namely, the *Gasteropoda*. Many species belonging to this class are terrestrial, but some (Limnæidæ) live in fresh water and have considerable economic significance. They serve as hosts for the intermediate forms of the Liver Fluke (*Distomum hepaticum*). *Limnæa truncatula* is the species which is chiefly concerned.

Although snails may readily be distinguished from slugs on account of their conspicuous shells, yet there is really very little difference between the two: slugs have rudimentary shells, which in some species are external but in others are concealed beneath the skin

The chief characteristics of the Gasteropoda are the head, the foot, and the respiratory apparatus.

The Head.—This is furnished with two pairs of tentacles or stalks, the longer and posterior pair bearing the eyes at their extremities. The mouth is furnished with two special structures for devouring vegetable tissue. In the upper part is a crescent-shaped jaw of a horny nature, whilst on the floor of the mouth is the so-called *radula*. This takes the form of a short, broad strip of tissue resting on a pad of cartilage. The upper surface of the radula is provided with a number of longitudinal rows of fine teeth, after the manner of a file. The radula can be moved backwards and forwards, and also slightly up and down. Plant tissue is held against the upper jaw and rasped away by the action of the radula.

The Foot.—This is a symmetrical structure on the ventral surface, and is provided with powerful muscles which enable the animal to progress. It secretes slime, which is also capable of sealing up the orifice of the shells of snails when necessary.

The Respiratory System.—Snails and slugs are air-breathing animals, and the gaseous exchanges necessary for respiration take place in the *mantle cavity*. The mantle is a fold of skin which arises from the dorsal surface of the body. It encloses between itself and the rest of the body a small air space, or mantle cavity, which communicates with the exterior through the pulmonary aperture, situated on the right side of the body immediately in advance of the shell. In snails the shell is conspicuous and large, consisting of several right-handed spirals; the viscera are stored within it.

Gasteropods have well-developed but asymmetrical nervous,

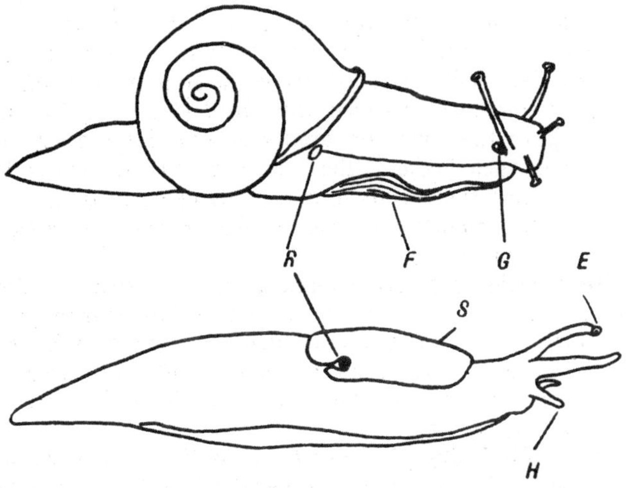

FIG. 149.—*A*. Edible Snail (*Helix pomatia*). *B*. Grey Field Slug (*Limax agrestis*). *E*. eye: *F*. foot: *G*. genital aperture: *H*. one of the short horns: *R*. opening to breathing, excretory, and digestive systems: *S*. shield.

alimentary, and excretory systems. The excretory pore and the posterior opening of the alimentary canal open close to the respiratory pore (Fig. 149).

Snails and slugs are hermaphrodite, having both male and female sex organs in one individual. The genital opening is on the right side of the body, immediately behind the head. Cross fertilization is general, however, and large eggs enclosed in a gelatinous layer of material, covered by a thin calcareous shell, are laid in batches in the soil, or beneath heaps of refuse. There is no metamorphosis, the young gasteropod after hatching resembling its parent externally in everything but size.

Two families of Gasteropoda are of economic importance:

FAM. HELICIDÆ.—**Snails.**—These are usually garden pests, the **Garden Snail** (*Helix aspersa*) being the commonest species. Occasionally the **Strawberry Snail** (*Hygromyia rufescens*), which is scarcely half an inch long, may cause considerable damage to strawberries. **The Wood Snail** (*Helix nemoralis*) is sometimes destructive to clover in the neighbourhood of thickets, etc.

Snails cannot be attacked in quite the same way as slugs, and hand-picking is often the best method of destroying them. They collect in numbers in any kind of suitable shelter, old walls, etc.

FAM. LIMACIDÆ.—**Slugs.**—The absence of a conspicuous spiral shell serves to distinguish slugs from snails, but all slugs have small, inconspicuous, external or internal shells of some kind. Many have small, flat shells concealed under the skin of the back. The region over this shell is called the *shield* (Fig. 149, *S*). Slugs, moreover, have the power of rapidly forming large quantities of slime all over their bodies: they can form two successive layers very rapidly, and these coats form a valuable protection.

The two most important species of slugs are the Grey Field Slug and the Black Field Slug.

The Grey Field Slug (*Agriolimax agrestis*) is either uniformly grey in colour, or flecked with brown and yellow markings. The animal lays its eggs in batches of about a dozen: they have been described as resembling grains of boiled sago. The young slug becomes sexually mature in about three months, and lives approximately eighteen months.

The Black Field Slug (*Arion hortensis*) is rather smaller than the preceding one. It is black on the dorsal side and yellow or orange underneath. Its slime is yellow or orange, whilst that of the Grey Slug is white and transparent. The Black Slug may lay up to seventy eggs in two or three bunches.

Nature of Injury caused by Slugs.—Slugs are most active in wet, mild weather, and are most frequent on heavy, damp soils. They feed chiefly at night. They attack cereals in the autumn and during a mild winter, often cutting off the shoots below ground. Later they attack and destroy the leaves. Clover seedlings are also attacked, and slugs are probably responsible for much of the loss of plant which is put down to other causes. Swedes, potatoes, peas, beans, and many garden crops are also damaged.

Control Measures.—Until recently slugs have been difficult to attack under field conditions, because they have been able to throw off such dressings as soot and lime in the copious slime which they excrete. It has been found, however, that copper sulphate has a paralysing effect on them, and is very deadly. It is recommended to mix one part of finely ground copper sulphate with 20 parts of kainit, and distribute the mixture at night at the rate of 3 cwt. per acre. It is necessary to carry out the work at night or in the very early morning, because it is essential that the mixture shall actually come in contact with the upper surface of the slug. The success of this method depends largely upon the evenness of distribution of the dressing, and the selection of a suitable time when the majority of the slugs are on the surface of the soil. It is particularly useful when cereals are being attacked, but may scorch broad-leaved plants. Excess of decaying vegetable matter in the soil tends to increase the slug population, and heavily infested fields should for a time receive artificial manure only. In the garden, trapping devices may be employed, and these are successful if regularly used. Young ducks are very effective in clearing out slugs and may be advantageously turned in among many crops.

METHODS OF COLLECTING, PRESERVING, AND EXAMINING SPECIMENS

THE agricultural entomologist needs very little apparatus, the most important desiderata being a good pair of eyes and an inquisitive disposition. For collecting purposes, however, the following articles will be found useful:

1. *Collecting boxes.*—These are cylindrical, and may or may not have glass tops: they are sold in ' nests ' of three or four, thus taking up very little room in the pocket.
2. *A killing bottle.*—A very satisfactory one can be made out of any convenient-sized, wide-necked bottle, by chopping up a leaf or two of the common Laurel and putting the pieces at the bottom. A piece of stiff paper, cut to the correct size and perforated with small holes, should be put into the bottle to keep specimens off the leaves. A killing bottle of this type is more satisfactory than one containing cyanide of potassium, which, besides being exceedingly dangerous, causes specimens to become very rigid and to lose their bright colours. Laurel leaves kill very efficiently, and keep specimens in a relaxed or softish condition, which is an essential preliminary to the ' setting ' stage. Beetles are best killed by being dropped for a second into boiling water.
3. *A collecting net.*—This is useful for Lepidoptera, Diptera and Hymenoptera. A collapsible one which goes into the pocket is to be preferred. A ' sweeping ' net for taking specimens from grass, weeds, etc., is also desirable.

For displaying insects to the best advantage some further apparatus is required, including entomological pins, setting-boards (for Lepidoptera, etc.), needles mounted in wooden handles, a good lens, and display cases. These last are expensive: good substitutes can be made out of shallow boxes lined with white paper.

Many insects have to be ' set ' before they can be added to a collection. In the case of Lepidoptera the relaxed insect is pinned through the thorax, with its abdomen resting in the trough of a setting-board. Its wings are then extended at right angles, and kept in position by straps of paper pinned over them. After drying for a week or two the wings remain extended when the straps are removed. (For full details of the methods of setting Lepidoptera consult the book on Moth Collecting mentioned in the Bibliography.)

Beetles are pinned through the base of the right wing cover, and the legs are extended in as natural a position as possible, gum being used to keep them in position if necessary. Flies and Hymenoptera look best mounted on discs of cardboard or transparent celluloid. Very small insects, such as Springtails, Lice and Fleas, are mounted in Canada Balsam, or some other medium, upon microscopic slides, and covered with glass slips.

Caterpillars can either be preserved in spirit or their empty skins can be blown out into their natural position and size. The operation is messy and requires a certain degree of skill. The animal is first plunged into alcohol to kill it, and then laid between sheets of thick blotting-paper. The operator next takes a pencil or cylindrical piece of wood, and starting at the posterior end, commences to force out the contents of the skin by gently rolling the pencil over the body, rolling always towards the anal end. When the skin is empty it is fixed on to the end of a blow-pipe, and is extended to its natural size whilst being held in an 'oven' made out of a tin situated above a burner. The skin eventually dries quite firm and may be wired to a straw or other support.

The anatomy of insects can only be studied through the dissection of various types. The Cockroach is an easy one to begin upon. The insect should be pinned out under water in a shallow vessel; small pie tins about three to four inches in diameter are suitable if a layer of melted paraffin wax is first poured into the bottom. For dissecting, a pair of fine scissors, a pair of fine forceps, one or two needles, and a sharp scalpel are required. Internal anatomy is studied by cutting away the dorsal plates with the scissors, and separating the fat body and tracheæ from the alimentary canal and other organs with needles and scalpel. Mouthparts can be examined most easily by cutting off the head and mounting it separately. A pair of fine forceps enables one to remove the structures, and mount them at once in water or balsam on a slide for examination with lens or microscope.

Students should consult the "Handbook of Instructions for Collectors," issued by the Natural History Department of the British Museum.

BIBLIOGRAPHY

General Entomology.

A General Text Book of Entomology. A. D. Imms. (Methuen.)
British Insects. Harold Bastin. (Methuen.)
British Plant Galls. E. W. Swanton. (Methuen.)
Insect Transformation. George H. Carpenter. (Methuen.)

Rhynchota.

The Plant Lice or Aphididæ of Great Britain. Vols. 1 and 2. F. V. Theobald. (Headley Bros.)
Hemiptera-Heteroptera. Saunders. (L. Reeve & Co.)

Lepidoptera.

Butterfly and Moth Collecting. Hodge. (C. Arthur Pearson, Ltd.)
The Moths of the British Isles (Series 1 and 2.) Richard South. (Warne & Co.)
The Butterflies of the British Isles. Richard South. (Warne & Co.)

Coleoptera.

British Coleoptera (6 volumes). Fowler. (L. Reeve & Co.)
Common British Beetles. Hall. (A. C. Black.)

Hymenoptera.

Wild Bees, Wasps and Ants. Saunders. (George Routledge & Sons, Ltd.)
Hymenoptera Aculeata. Saunders. (L. Reeve & Co.)
British Hymenoptera. Buckhurst, Staniland & Watson. (Arnold.)
A Monograph of the British Phytophagous Hymenoptera. Cameron. (Ray Society.)

Diptera.

British Flies—Syrphidæ. Verral.
British Flies—Stratiomyidæ. Verral.
Typical Flies (3 vols.). Pearce. (Cambridge University Press.)

Economic Entomology.

The Principles of Insect Control. Wardle & Buckle. (Manchester University Press.)
Insect Pests of Fruit. Theobald.
Agricultural Zoology. Theobald. (W. Blackwood & Sons.)
Insect Pests and Fungus Diseases. P. J. Fryer. (Cambridge University Press.)
Agricultural Parasitology. Walton & Wright. (Sidgwick & Jackson.)
Successful Spraying. P. J. Fryer. (Ernest Benn, Ltd.)

See also the publications of the Ministry of Agriculture, The British Museum (Natural History), "The Review of Applied Entomology," "The Journal of Economic Entomology" (Agricultural Section), "The Bulletin of Entomological Research," "The Annals of Applied Biology," and the reports of consulting entomologists in the Journals of the Royal Agricultural Society of England, and the Highland and Agricultural Society of Scotland.

INDEX